Japanese Wooden Boatbuilding

The Cormorant Fishing Boat

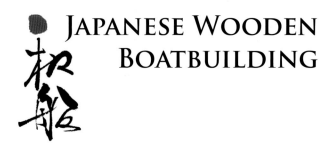

JAPANESE WOODEN
BOATBUILDING

The Cormorant Fishing Boat

A Japanese Craftsman's Methods

Douglas Brooks

Floating World Editions

Published by Floating World Editions, Inc.
26 Jack Corner Road, Warren, CT 06777

ISBN 978-1-953225-01-6

Author and publisher gratefully acknowledge the generous and continuing support of The Freeman Foundation, the Asian Cultural Council, Gifu Academy of Forestry Science and Culture, the Tokyo National Research Institute for Cultural Properties, and Georg and Franny Hinteregger.

To Catherine,
For always being there.

Also by Douglas Brooks

Building the Nagara River Ubune with Boatbuilder Nasu Seiichi (Funadaikou Nasu Seiichi to Nagaragawa no Ubune o Tsukuru), translated by Chieko Wales, Department of Intangible Cultural Heritage, Tokyo National Research Institute for Cultural Properties, 2020 (text in Japanese).

Japanese Wooden Boatbuilding, Floating World Editions, Warren, Connecticut, USA, 2015 (text in English).

Sabani: Building an Okinawan Fishing Boat, translated and edited by Koji Matano, Wooden Boat Center, Takashima, Shiga, Japan, 2014 (text in Japanese).

Ayubune, museum monograph written with Riko Okuyama, Mizunoki Museum of Art, Kameoka, Japan, 2014 (text in Japanese and English).

Construction Record of Sabani, the Traditional Okinawan Fishing Boat (Okinawa no Dentouteki Gyosen Sabani Kenzou Kiroku), Museum of Maritime Science, Tokyo, Japan, 2011 (text in Japanese and English).

"Ships of the Japanese Coastal Trade," in *Shipwrights Annual*, Annova Books, United Kingdom, 2010 (text in English).

"Bezaisen: Japan's Coastal Sailing Traders," in *Sailing Into The Past*, Seaforth Publishing, United Kingdom, 2009 (text in English).

The Tub Boats of Sado Island: A Japanese Craftsman's Methods (Shokunin no Gihou: Sado no Taraibune), translated by Chieko Wales, Kodo Cultural Foundation, Sado Island, Japan, 2003 (text in English and Japanese).

FORTHCOMING IN THIS SERIES

Isobune, small inshore fishing boat, Sanriku coast

River boats, Niigata Prefecture

Tenmasen, work boat, Himi, Toyama

Contents

Preface and Acknowledgments   9

Wooden Boats Built in Japan by the Author   11

The History and Magic of Ukai by Richard J. King   13

The Cormorant Fishing Boat, Ubune   19

Building the Ubune   39

Launching   67

Conclusion   69

Glossary   73

The Gifu City Shipyard and Museum Boats   75

Other Cormorant Fishing Boats   79

About the Author   82

Preface and Acknowledgments

My research involves working alongside traditional wooden boatbuilders in Japan, recording as much of their methods as I can, and creating measured drawings of the boats along with a narrative describing these techniques. All drawings and photographs in this book are my own unless otherwise noted. In Japan the craft of boatbuilding is infused with secrecy, and many of the craftspeople I have met built boats without any written record or drawings whatsoever, relying upon memory alone. Working from memory has protected secrets, but with no one to pass these secrets on to and little documentation by researchers, an enormous body of knowledge is at risk of being lost.

This was my seventh opportunity working with a craftsperson documenting their techniques. My first apprenticeship was in 1996, when I worked with the last active builder of tub boats (*taraibune*) on Sado Island, Niigata. He and I built one tub boat together and over the years I have built six more. In 2000 I was invited by one of the last boatbuilders of Urayasu, Chiba, to build a traditional Tokyo Bay seaweed boat called a *bekabune*. In 2002–2003 I lived in Japan for a year and built three boats: two with a boat builder from Sumida-ku, Tokyo, and one with a boat builder in Aomori Prefecture. In 2009 I studied with one of the last three traditional boatbuilders in Okinawa, building an eight-meter *sabani*, a type of fishing boat. In 2015 I traveled to Iwate to work with the last active boatbuilder in Sanriku, building an *isobune,* the most common wooden boat of the region that was devastated by the 2011 tsunami. In 2019 I built a river boat with the last boatbuilder of Niigata City, followed by a small fishing boat in Himi, Toyama, working with the last boatbuilder of that region. I have also built boats as exhibitions at museums and arts festivals across Japan and the U.S.

In *Japanese Wooden Boatbuilding* I documented the work of five traditional craftsmen and their boats. Although it is not necessary to have read that work to understand the present volume, I urge those with a deeper interest in the subject to do so. For those who have, I apologize for the repetition they will inevitably encounter here, although readers of both volumes will notice very significant differences in the construction of this boat compared to others I have described.

In the course of twenty-five trips to Japan I have also met and interviewed over fifty boatbuilders from all of Japan's main islands. Collectively these men taught only about half a dozen apprentices. The youngest boatbuilder I know from this generation is now in his late seventies. These men all epitomize a story sadly seen across a wide range of traditional crafts in Japan: the practitioners are elderly and traditions formerly nurtured for generations by the master-apprentice relationship are now disappearing. Mr. Seiichi Nasu, with whom I built the boat documented here, is the rare boatbuilder from this era who had taught an apprentice. None of my first six teachers had done so; I had been their first.

The goal of this book is to document as completely as possible the design and construction techniques employed by Mr. Nasu for building the *ubune* (more formally called *ukaibune*) a special river craft used for cormorant fishing, *ukai*. The design and construction of Japan's river boats differ greatly from the much more prevalent sea boats, and importantly, Nasu is the first river boat builder whose work I have documented. As the reader will see, Nasu's work is even more specialized, given the particular demands of cormorant fishing, and in this text I will occasionally offer comparisons to other styles and techniques to provide perspective on how building ubune fits within the broader context of traditional boats.

I began discussing ways of documenting how ubune are built with Mr. Masashi Kutsuwa, director of the woodworking program at Gifu Academy of Forestry Science and Culture in Mino. Kutsuwa's program is committed not only to training woodworkers, but to actively encourage them to study and revive local traditional crafts. He knew Nasu well and recognized the enormous significance of his work and the iconic status of the ubune in Gifu and Japan. In February of 2016 Kutsuwa and I went to see Nasu's former apprentice, Mr. Hiroshi Tajiri, and we broached the idea of my working with him. Kutsuwa continued to talk to Tajiri after I returned to America but ultimately the arrangement fell through. Amazingly, when Nasu heard our project was in jeopardy, he stepped forward and agreed to teach me. He made it clear to us he was not strong enough to perform much physical work, but he would supervise the construction and lay down all the measurements. Thanks to Nasu's remarkable generosity and commitment, fifteen years after our first meeting I was finally building a boat with him.

I was joined in the workshop by Mr. Marc Bauer, a naval architect from San Francisco, Mr. Satoshi Koyama, a student at Gifu Academy, and Mr. Hideyaki Goto, a *sendousan* (boatman) for the Gifu City tourist boats. Masashi Kutsuwa was available most days to assist in occasional translation, as was Ms. Migiwa Imaishi, a researcher from Tobunken, the Tokyo National Research Institute for Cultural Properties, our partner in this research.

In Japan apprentices learn by observation rather than by direct instruction, taking on menial tasks and painstakingly developing skills while watching their masters. All work takes place in strict silence, a code most of my previous teachers enforced. In our case we all had varying degrees of boatbuilding experience and Nasu gave us instructions and answered questions whenever we needed him. He laid down all the dimensions and demonstrated techniques while we did the physical work. We did our best to record Nasu's methods as carefully as possible; however, I take full responsibility for any errors or omissions in this text.

Funding for this project came from The Freeman Foundation of Honolulu, Hawaii, and from Tobunken, which published a Japanese version of this material in March of 2020, available for digital download from Tobunken's website.

The Gifu Academy of Forestry Science and Culture provided a temporary workshop and ongoing organizational support. Mr. Kentaro Hiraku, who runs a business giving customers tours of the Nagara River, agreed to purchase our boat, providing us with additional funding. We also received funding from Georg and Franny Hinteregger.

Learning about building the ubune naturally draws one into the fascinating history and lore of cormorant fishing. An introduction to the "History and Magic of Ukai" has graciously been provided by marine natural historian Richard J. King, excerpted from his article in *Maritime Life and Traditions*. As well, a 2018 research grant from the Asian Cultural Council of New York allowed me to seek out and study other types of cormorant fishing boats found in Japan and China. Information on those boats and photos can be found in the Appendices.

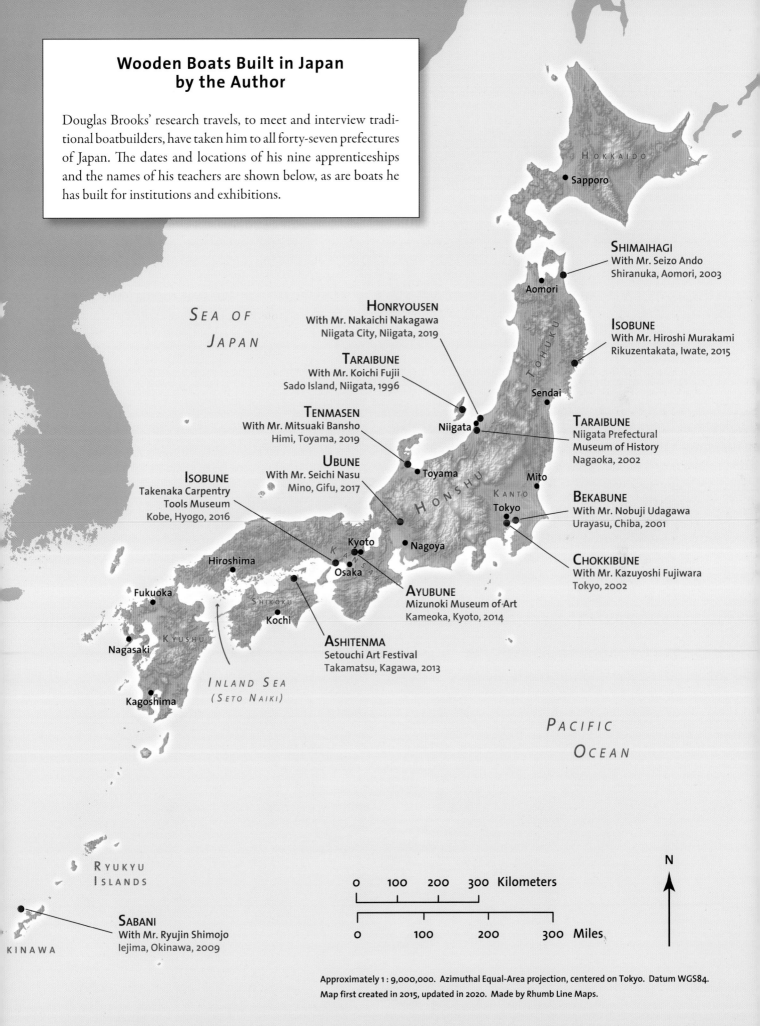

Wooden Boats Built in Japan
by the Author

Douglas Brooks' research travels, to meet and interview traditional boatbuilders, have taken him to all forty-seven prefectures of Japan. The dates and locations of his nine apprenticeships and the names of his teachers are shown below, as are boats he has built for institutions and exhibitions.

SEA OF JAPAN

HOKKAIDO

● Sapporo

SHIMAIHAGI
With Mr. Seizo Ando
Shiranuka, Aomori, 2003

● Aomori

ISOBUNE
With Mr. Hiroshi Murakami
Rikuzentakata, Iwate, 2015

HONRYOUSEN
With Mr. Nakaichi Nakagawa
Niigata City, Niigata, 2019

TARAIBUNE
With Mr. Koichi Fujii
Sado Island, Niigata, 1996

TOHUKU

● Sendai

TENMASEN
With Mr. Mitsuaki Bansho
Himi, Toyama, 2019

Niigata ●

TARAIBUNE
Niigata Prefectural
Museum of History
Nagaoka, 2002

UBUNE
With Mr. Seichi Nasu
Mino, Gifu, 2017

● Toyama

HONSHU

● Mito

ISOBUNE
Takenaka Carpentry
Tools Museum
Kobe, Hyogo, 2016

KANTO

Tokyo ●

BEKABUNE
With Mr. Nobuji Udagawa
Urayasu, Chiba, 2001

Kyoto ●

KANSAI

● Nagoya

● Osaka

CHOKKIBUNE
With Mr. Kazuyoshi Fujiwara
Tokyo, 2002

● Hiroshima

AYUBUNE
Mizunoki Museum of Art
Kameoka, Kyoto, 2014

● Fukuoka

SHIKOKU

● Kochi

KYUSHU

ASHITENMA
Setouchi Art Festival
Takamatsu, Kagawa, 2013

● Nagasaki

INLAND SEA
(SETO NAIKI)

● Kagoshima

PACIFIC OCEAN

RYUKYU ISLANDS

N

0   100   200   300   Kilometers

0   100   200   300   Miles

SABANI
With Mr. Ryujin Shimojo
Iejima, Okinawa, 2009

KINAWA

Approximately 1 : 9,000,000. Azimuthal Equal-Area projection, centered on Tokyo. Datum WGS84.
Map first created in 2015, updated in 2020. Made by Rhumb Line Maps.

An usho fishing in Seki, tending the leashes of his cormorants, one of which is visible silhouetted by the reflection of the fire. Cormorant fishing is done at night, and the light from the wood-fired brazier attracts the fish.

The History and Magic of Ukai

Richard J. King

You would be hard-pressed to find anything today that feels as ancient, strange, and beautiful as people fishing with cormorants aboard traditional wooden boats on a dark summer evening on a river in Japan, an image that hearkens back at least to the early seventh century in parts of eastern Asia. People have fished with cormorants in China, Japan, India, and Vietnam, and possibly even in ancient Egypt and Peru. Most historians believe the practice originated in China or Japan or perhaps evolved independently in each of those nations. In China the practice has been conducted since at least the Song Dynasty (960–1298); some archaeologists believe they have evidence of the method in use as far back as 317 BC. Eventually, fishing with cormorants became widespread as a subsistence fishery over a great part of China, but now it is conducted predominantly in the Lower Yangtze basin.

In Japan, the first known written record of cormorant fishing seems to have come from a Chinese envoy visiting in AD 607. It is written in the Chinese *Sui shu,* completed in 636: "In Japan they suspend small rings from the necks of cormorants, and have them dive into the water to catch fish, and that they can catch over a hundred a day." The Japanese term for the activity, *ukai,* literally "cormorant keeping," is mentioned in the *Manyoshu,* the earliest book of Japanese poems, compiled in the eighth century. The *Kojiki (Chronicles of Old Matters)* and the *Nihon shoki (Chronicles of Japan,* from the earliest times to AD 697) are Japanese historical records that were also completed in the eighth century. These two volumes suggest that ukai had been taking place in Japan since even before the seventh century.

Historians have found other ancient mentions of ukai in Japan. In AD 702 census records in a town near Gifu City

show a person with the name Ukaibe or Uyobe, suggesting this could have been an *usho* (the most common modern term for the practitioners), since men would take the name of their trade. Census records during the Enji Period (901-22) reference seven houses of cormorant fishermen on the Nagara River. The fishermen sent their catch of *ayu, Plecoglossus altillelis,* a salmonid commonly called sweetfish, to the imperial court through the governor of the province, Toshihito Fujiwara. There is also reference to ukai in the *Tale of Genji,* considered one of the oldest novels in the world, written in the early eleventh century and credited to Lady Murasaki Shikibu. Records from the twelfth century show an increase to twenty-one usho houses in seven villages along the Nagara, and an account in 1190 says that cormorant fishermen delivered "sugared sweetfish and rice" to officials.

In the sixteenth century there is record of a female usho fishing on the Nagara River. Her name was Aka, and she was

A box of newly caught ayu, a freshwater fish popular throughout Japan, some showing a prominent puncture wound made by a cormorant's beak.

said not only to be very skilled, but also to fish with twelve cormorants, perhaps one of the first to do so. The Japanese scholar S. Ikenoya wrote: "This lady of old is regarded as the mother of ukai."

In 1564 Nobunaga Oda, a warlord and major figure in the history of Japan, helped further ukai on the Nagara River. He treated cormorant fishermen with respect, granting them the honorable position of usho, and giving them the same rank as falconers. Oda granted the fish and ukai special status at Gifu Castle, where he briefly lived. He gave the usho a bag of rice each month and had a new boat built for them every four or five years. In 1568 Oda invited the servant of another lord in the region to come and see ukai as a spectacle. Historians mark this as perhaps the first instance of when ukai was seen as something to watch for entertainment. In the following years ukai started to become a court event, the magical night vision of this practice commonly viewed by nobility.

During the time of the Tokugawa Shogunate, beginning in the seventeenth century, ayu became a popular sushi fish for the military, and ukai came under direct control of a feudal clan. Records show that fourteen usho on the Nagara River and seven on the Oze received money to purchase rice in exchange for ayu sushi packed in barrels and sent to the Shogunate's castle in Edo, while ukai was protected by the outlawing of dams, weirs, and net-fishing on the Nagara River, anywhere near where cormorant fishing was taking place.

In the eighteenth century, the scholar Norinaga Motoori wrote of ukai in Gifu, of that night spectacle on the river:

Nowhere but in the Nagara can we see
That antique sight of cormorant fishing,
So picturesque and impressive,
Bonfires reflected in the water rushing.

The Imperial Restoration of 1868 ended the Tokugawa Shogunate and its protective policies, which meant a temporary cessation of the stewardship of ukai, but, fortunately, a decade later and again in 1880, Emperor Meiji came to see ukai at Gifu. In 1890 the emperor issued a decree declaring that cormorant fishing was protected by the Imperial Household Agency. He set aside three areas along the Nagara River as imperial fishing places.

Today, cormorant fishing is conducted entirely for the benefit of tourists and takes place in about a dozen locations throughout western Japan. The boats are all of different designs, except in Seki, just upstream from Gifu, where the usho generally buy used boats from the Gifu fishermen, and in Iwakuni, Arashiyama, and Uji, where Gifu boats have also been sold. In one community the river is too shallow for boats, so the usho wade through the stream leading their cormorants. The usho of Gifu and Seki remain by far the strictest adherents to traditions.

As early as 1924 the City of Gifu decided also to sponsor ukai, and today it is subsidized by the city government. Now, around the city of Gifu, artisans of all kinds have been commissioned to create cormorants and ukai scenes on sewer caps, public sculptures, and as cartoon mascots for the post office and on the sides of buses. There is an entire museum devoted to the history of ukai. The architects of the tallest building in Gifu overtly drew upon the boat-building techniques of the ukaibune when designing interior spaces. The Imperial Household donates a small amount towards the usho salaries, but the city covers the rest. Periodic surveys have shown that tourist revenue from ukai more than makes up for the investment.

From early summer to late fall in Gifu City, cormorant fishermen take to their boats every night, unless it is a full moon or the water is dangerously high or too cloudy due to rain or sediment. The usho fish the river, which runs through the middle of the city in a valley winding between steep hills. Gifu Castle, a twentieth-century reconstruction of the fortress that dates back to 1201, overlooks this portion of the river from the tip of Mount Kinka. Usho on the Nagara River fish almost exclusively for ayu, a fish found in the streams, lakes, and coastal waters of Asia that can grow up to a foot long, although those caught by the cormorants during ukai season are normally much smaller.

As it has been for centuries in part, ukai today is a show. The spectators—both foreign visitors and Japanese people from other parts of the country—come in buses and pile onto the dock where one of the fishing masters wears a microphone and gives a talk and demonstration with a couple of his birds. (In Gifu, the usho are all men, as they have been for generations, but today in Uji, a city south of Kyoto, there are two female usho.) The usho explains his costume and the history and process of ukai. He shows how he and an assistant tie a string at the base of the bird's neck. He feeds fish to the cormorant to demonstrate how its long neck expands with the food in its gullet. To the amazement of the spectators, the usho then makes the bird regurgitate the fish.

These are Japanese cormorants (*Phalacrocorax capillatus*), nearly goose-sized beautiful black seabirds with long sharply-curved beaks that the usho have captured full-grown from the northern sea coasts and domesticated them here in

Cormorants in their coop during their "time off." Note the baskets in which they will be transported to their boats for fishing.

the river over many months. These cormorants have emerald green eyes, and to swim expertly underwater they have wide webbed feet and water-absorbent feathers that allow them to be less buoyant below the surface. The capture and keeping of these wild birds for fishing is not without some ethical debates, especially in recent decades, but concern for these animals has a centuries-long history.

After the introduction, the usho, his assistants, and the cormorants (four to a basket) return upriver to prepare for the evening's fishing. The men wait for darkness and draw lots to determine boat position. Meanwhile, some spectators get ready to watch ukai from the rocks along the shore, but the more popular way to observe is from a dinner boat. Following the usho's demonstration, men, women, and children take off their shoes and go aboard small barges. Gifu City employees manage dozens of dinner boats, some powered by engines, but more often driven simply by long poles and the river's current.

The spectators are floated upriver and under the bridge, where they watch the sun go down behind the mountains. They eat and drink, anticipating the show. They listen to recorded traditional Japanese music and watch entertainers in traditional Japanese dress perform on a boat that navigates between them. On some evenings there are fireworks, or the city closes the roads along the shore to remove the noise and lights of cars. The dinner-boat crews decorate the vessels with paper lanterns, and some passengers light sparklers, particularly if they have children with them—or if they've had a few drinks.

The first thing the spectators see of ukai is a glow beyond the bend of the river. The light is from the usho's fires, and soon the six boats themselves appear. Everything goes silent and the river is transformed. It is difficult to remember that you are in the middle of a major city and that the shores are lined with tall buildings. The spectators hear the oars through the water and the splashing of the birds. The dinner-boat crews maneuver their vessels in a choreographed dance around the fishermen, staying out of the way while providing every passenger a close-up view.

Ukai is spellbinding, and the usho of Gifu and Seki pride themselves on the historical authenticity of their boats and their work clothes. The usho stands at the bow to tend his fire, the *kagaribi*, which hangs out forward in the *kagari*, an iron brazier or cresset suspended from a wooden pole that fits into a slot in the bow. The base of this pole, wrapped with leaves of Rose of Sharon, provides lubrication as he pivots it from side to side, wherever the usho wishes to cast the light. He fuels the flames with slats of pine, *matsuwariki,* until the fire crackles and reflects off the water, attracting the coveted fish. Embers pop and sizzle, flying down into the water or back into the boat. To protect himself, the usho wears a uniform that only adds to his appearance as a sorcerer. He wears a dark *kazaorieboshi,* a linen cloth that he has wrapped around his head in order to shield his hair, and a *ryofuku,* a full-length black or dark blue cotton kimono, which he has folded up under his crucial straw skirt, the *koshimino* that extends below the knee and serves to repel water, insulate, and guard his legs from the wings and feet of the birds. Even then an usho still gets small burn holes in his kimono and his *muneate*, an additional layer worn over his chest. On his feet he wears *ashinaka*, straw sandals with only half a sole so that his heels meet the deck to keep from slipping.

While tending the fire, the usho manages a flock of twelve cormorants swimming in the river. He holds their leashes, the *tanawa*, in one hand while with the other he adjusts each string—plucking, overlapping, tugging, slacking—as the birds dive, fish, and return to the surface with their gullets full. While still managing the others, he pulls individual cormorants back to the boat, picks them up, squeezes the fish out of their necks into a basket, and then returns them to the water. The fire illuminates the cormorants' wet heads and the patch of white and yellow on their faces. When not diving, the birds swim ahead of the boat beneath the flame, as if pulling the craft forward, like swimming reindeer guiding a dark blue Santa Claus.

The leashes fan out from the usho's hand, with the rope tied loosely enough around the bird's neck so that movement is not restricted, but firmly enough to stop the passage of larger fish down into its stomach. The cormorants will eat small fish while they are diving and could eat the larger ayu if they really wanted, but the usho believe there is an "understanding": the birds drink the fish oil, their favorite part, then allow the usho to take their catch, since they know they will get whole fish later. The material and lay of the leashes resist tangling and fraying. If a leash is pulled hard it won't break, but if twisted sharply against the lay of the line, it will snap. In this way the usho can free a bird if its leash gets caught under a rock. Between the leash and the neck loop is a thin stick of stiff plastic, which hangs parallel to the cormorant's back. This helps to keep the line out of the bird's way and assists the usho in his handling of the cormorant. Until the middle of the twentieth century, this stick was made of whale baleen.

In each boat the usho has two or three assistants. At the stern, steering with either an oar or a long bamboo pole is the *tomonori*. Amidships is the *nakanori*, who helps to navigate with a smaller oar and also to manage the incoming fish and the cormorants in their baskets. The third assistant is the optional *nakauzukai* who sometimes sends six additional cormorants into the river and manages them from the waist of the boat.

The demonstration of ukai lasts less than one hour. The climax is the *sougarami*, when all six boats form a line across the river and drive the fish. The usho shout. The nakanori bang their oars against the hulls. Besides the intrusion of a few camera flashes, the atmosphere and experience feel truly

from a centuries past. After the sougarami the fishermen drift their boats over to the rocky beach to let the cormorants go ashore. The birds croak and walk around, some standing with their wings spread to dry them. The usho remove the cormorants' leashes and without irony each man rewards his birds with fish.

Meanwhile, the dinner boats return under the bridge and to the dock. As at the end of a sad movie, most of the spectators appear subdued and pensive. Enhancing this somber impression, a Japanese version of *Auld Lang Syne* is played through giant speakers. The old Scottish song echoes across the river and is surprisingly moving. If you think this nostalgic emotion sounds like nonsense, even before the use of the song, consider that the Japanese poet, Matsuo Basho wrote of ukai in 1688 (as translated by D. Barnhill): "The cormorant fishing of Nagara River at Sho in Gifu is famous throughout the country and it is as fascinating as the accounts. Without wisdom and talent, I cannot possibly exhaust the scene in words, but I long to show it to those whose heart understands." Basho wrote this haiku:

omoshirote	Exciting to see
yagate kanashiki	but soon after comes sadness
ubune kana	the cormorant boats

In today's Japan ukai is a cultural display, a form of entertainment, and also a way to preserve historical skills and practices; it is not unlike the chantymen of Western maritime museums who entertain the public while serving to maintain their culture's bank of maritime techniques and sea stories. The first fish of the season are still presented to the Emperor from the Governor of Gifu Prefecture and the Mayor of Gifu City. According to official Gifu City tourism brochures, 122 tools of ukai are listed as "Important Tangible Folk Custom Cultural Assets of the Nation." Although ukai has not been conducted solely as a fishery business in Japan for many decades, it remains a valued and valuable cultural asset, a popular and enduring form of entertainment for more than four hundred years.

Excerpted and revised from "Mr. Yamashita," an article by Richard J. King appearing in the Summer 2006 issue of Maritime Life and Traditions *(No. 31).* Dr. King is the author of four books of nonfiction, including The Devil's Cormorant: A Natural History and Ahab's Rolling Sea: A Natural History of Moby-Dick.

Tourist boats gather to view cormorant fishing on the Hiji River in Ozu, Ehime Prefecture, circa 1960. The activity was revived after World War II as a way to promote tourism. Photo courtesy Kita Management.

A Seki boatman waits for darkness to fall before guiding his boat full of tourists into the stream to watch the cormorant fishing.

The Cormorant Fishing Boat, Ubune

Introduction

My first day working with Seiichi Nasu in 2017, building a cormorant fishing boat, he turned to me and said, "You know, we've been talking about this for twenty years." I first learned about Nasu in the mid-1990s from a newspaper article, but according to my notes we didn't actually meet until 2003. At the time I was traveling around Japan trying to find as many boatbuilders as I could, photographing their work and interviewing them for my research. Meeting Nasu it was obvious he had led a remarkable life. The boat for which he was known—the *ubune*[1]—was central to an iconic fishery dating back 1,300 years on the Nagara River. At the time he was the only builder, having taken over from a man on the Kiso River named Shoji Mishima twenty years earlier. In his younger years Nasu and his father had built hundreds of smaller fishing boats for customers on the Nagara and Kiso Rivers. When we first met he had an ubune in his shop, almost finished. When I asked him about drawings he said he had none: all the dimensions were memorized, and he showed me a few wooden patterns I had no idea how to use. I asked about the possibility of apprenticing with him and he shrugged off the suggestion, saying he wasn't sure when he would have another commission.

A few years later I visited Nasu a second time, and again I asked him if it was possible to study with him. He said he was getting older and building an ubune now took many months. After a few days of work he was forced to rest for several days. His market consisted primarily of the six *usho,*

or cormorant fishermen, of Gifu City, who worked every night for six months of the year demonstrating their techniques for tourists who crowd spectator boats. This activity, as described in detail by Richard J. King earlier, takes place on the Nagara River in the center of the city and may be Gifu's largest single tourist attraction. In recent years an average of 100,000 people annually come to watch cormorant fishing; however, this is down from a peak of over 300,000 visitors in 1973.

The fishermen, all men who have inherited their roles, are paid a stipend from the Gifu Tourist Association as well as the Imperial Household Agency, the bureaucracy that supports the Emperor and his family. In fact, usho are required every year to stage eight evenings of fishing specifically for members of the Imperial Household and send the catch to the Imperial Palace. Gifu City also operates a shipyard, building and maintaining the fleet of dozens of spectator boats. A smaller tourist operation exists nearby in Seki City, as well as in about a dozen other locations in western Japan, though the Gifu fishermen are by far the most well known and the only ones to receive Imperial support.

In addition to the six Gifu usho, three fisherman in Seki may sometimes order a new boat but often buy used boats from their compatriots in Gifu. The Seki fishermen receive no Imperial stipend and host a much smaller audience, about 7,500 tourists a year. Ubune have relatively short useful life spans, generally ten to fifteen years. Thus over the previous twenty years Nasu had basically built replace-

---

[1] The *ubune* 鵜船, literally "cormorant boat," is also called ukaibune 鵜飼船, or "cormorant fishing boat." Cormorant fishermen are called *usho* 鵜匠.

Boatbuilder Seiichi Nasu discussing the project with his daughter, Setsuko.

Ubune tied along the riverbank in Gifu City. These boats were built in the last ten years by a craftsperson named Hiroshi Tajiri of Gujo, who studied with Nasu almost forty years ago.

ments—seven ubune in all—for all of the boats Mishima had built before him.

On my earlier visit I had given Nasu a copy of my first book documenting the history, design, and construction of the tub boats of Sado Island. He told me he had passed the book along to Mr. Hiroshi Tajiri, the man who apprenticed with him over twenty years earlier, and who had been making bathtubs and buckets in the intervening years. Within a few years of this meeting Nasu began passing on his clients to this craftsman, who built all of the ubune currently being used by usho in Gifu City and all but one of the boats used in Seki City, about ten boats in all, slowly replacing all the boats Nasu built. The Seki boat may be the only ubune built by Nasu still in use. The boat we were building together—the subject of this book—was his eighth ubune.

In addition to the ubune in use today on the Nagara River there are two historic ubune in the collection of the Gifu City Museum of History. One of these boats was built by Gosaku Ando. A 1985 documentary film follows Ando building this boat, his last.

Mr. Seiichi Nasu

At the time we worked together Nasu was eighty-five years old. He was born July 21, 1931, and had lived in Mino all of his life. His home and workshop were in a small hamlet upriver from Mino City, where the river suddenly narrows, winding its way through a small mountain pass. His father was a boatbuilder, having learned the craft from a man named Toyoda in Hodojima, Seki. Nasu began his apprenticeship with his father at age seventeen and said his training lasted about six years. He and his father worked together for over twenty years, building boats of various sizes for customers on the Nagara and Kiso Rivers, with a few boats built for the Ibi River. He told us his boats ranged from three meters to twelve meters in length, but an important distinction is that Nasu, like many boatbuilders in Japan, describes the "length" of a boat as the length of the bottom only. He said in his younger days it was not uncommon for him to work from 5 am to 10 pm.

Today these rivers support the famous ukai fishing tourist industry as well as sport fishing, but in Nasu's younger days *ayu* fishing was an important commercial enterprise. Ayu are a small, fresh-water fish found throughout Japan. Normally translated as "sweetfish," ayu have white flesh and are considered a delicacy. The spring season of ayu is eagerly anticipated in Japan. In some regions massive traps are used to capture fish and in other places temporary restaurants are erected on the river banks to accommodate the demand for fresh fish, reminiscent of the "shad bakes" I remember growing up on the Connecticut River.

Nasu mentioned building *tabune* for use in the rice paddies as well as *umabune*, a type of ferry capable of carrying a horse or cow. Fishing boats came in a variety of sizes, one he called a *ryosen*[2] which came in two types: one

---

[2]   While ryosen typically means "fishing boat" in this case *ryo* reflects the alternate meaning "both," as both ends of this type of boat are pointed ("sharp" in nautical terminology).

for net fishing, called *amibune,* and one for rod fishing, called *tsuribune.* Other boats included *yotsunori* and *ishibune.*[3] Each type represents a distinct design, though ryosen share many of the design details of the ubune. Nasu himself still owns a 16.5 and a 20.5 *shaku*[4] yotsunori. He says he has built almost six hundred boats.

Nasu stressed he and his father had excellent techniques but just as important, they listened to their customers' needs to customize designs. This gave them a reputation that attracted additional business. Nasu said if a client asked his father to build a boat quickly his father would tell him to find another boatbuilder. He said he felt the same about his work, and added this made him worry about our project, because we were operating under a tight schedule. In later years, as other boatbuilders retired, their customers came to him. By the late 1980s he was the last remaining boatbuilder in the region. Building new ubune actually started late in his career.

Nasu's detailed understanding of boats also came from part-time work as a boatman. Starting at age ten he worked first on a cargo boat and then on fishing boats. These were net boats catching ayu and the fishing was seasonal, from May to October. The fishing took place at night starting about 8 pm and often lasting until midnight. Fishermen would set their nets in the river then head upstream, returning downstream, driving the fish into the nets. He told us when conditions were right he would work for fishermen up to four nights a week. This could be very lucrative work, as Nasu said in 1950 he was charging 12,000 to 15,000 yen for a new fishing boat and in good conditions a fisherman could earn that much money in two nights! However, in leaner times, he said fishermen might take as long as three years to pay for a boat, giving him payments twice a year at the late-summer Obon Festival and New Years. Nasu said he sometimes had to pay his lumber broker the same way.

Amazingly, Nasu told us he worked as a boatman until age seventy, a sixty-year career! This is testament to the economic environment in which craftspeople like Nasu worked; long hours in the workshop as well as additional work outside of their craft in order to make a living.

Nasu and his father sometimes traveled for work. When building a new boat a customer would prepare the materials while the boatbuilder would bring tools and nails. The customer would house and feed the craftspeople, providing a workspace, but Nasu said customers did not participate in the building of their boats. (Traveling for work was common for rural craftspeople across many trades. My teachers in Sado, Aomori, and Iwate described similar arrangements.)

Nasu told a story of how he and his father sometimes delivered boats via a cart, which they pulled themselves as far as fifteen kilometers. Once the boat was delivered, Nasu's father negotiated to buy the customer's old boat. If successful, he and his father would load their cart onto the old boat and return home by river. They would then repair the old boat and re-sell it. This kind of arduous labor came to an end in 1960 when Nasu got his driver's license and bought a truck.

In his father's era fishermen ranged over a much longer extent of the rivers, traveling by boat seasonally. In his father's time Nasu said fishing took place as far away as Gujo, thirty miles upstream, so the usho had to tow their boats far upriver. In the winter, usho would take their boats downriver so their cormorants could feed. This year-round use of boats and travel greatly diminished their life and Nasu estimated in the old days ubune lasted only three to four years.

Shop and Tools

Our ubune was built in a temporary workshop erected by students at Gifu Academy of Forestry Science and Culture right outside the school's woodworking shop. Nasu had lost his shop in a major flood, and his replacement shop was too small to build an ubune. While the school's workshop had a complete set of large stationary power tools we only used them incidentally. He and his father built boats with just hand tools. Though Nasu always remembers the house having an electric light he didn't buy a power tool until

Nasu's *sumitsubo,* or ink line. Most craftspeople fill these with with black ink but Nasu uses *bengara,* a red coating normally used as a wood finish.

---

3 Yotsunori are fishing boats with transom (flat) sterns. Ishibune are boats meant to carry stones.

4 A chart explaining the traditional measuring system can be found on p. 24.

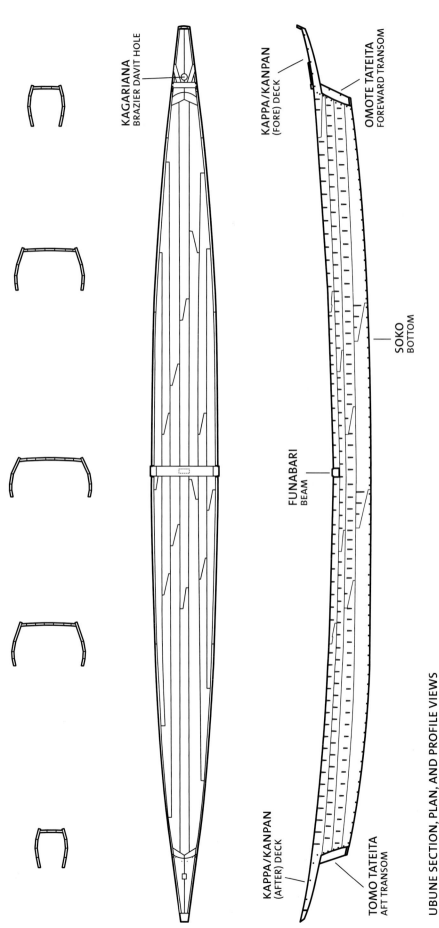

KAGARIANA
BRAZIER DAVIT HOLE

KAPPA/KANPAN
(FORE) DECK

OMOTE TATEITA
FOREWARD TRANSOM

SOKO
BOTTOM

FUNABARI
BEAM

KAPPA/KANPAN
(AFTER) DECK

TOMO TATEITA
AFT TRANSOM

UBUNE SECTION, PLAN, AND PROFILE VIEWS

From top to bottom, the section, plan, and profile views of the boat built under the direction of Mr. Seiichi Nasu in 2017. The ubune is distinguished by two long decks fore and aft: the fire brazier swings from a davit mounted in the foredeck while the after deck provides a resting place for the cormorants. The smooth (not overlapping) planking is very rare in Japanese boatbuilding. The relatively narrow bottom planks are meant to ease repair when the bottom is damaged running rapids in the Nagara River. Drawing by Ms. Migiwa Imaishi.

1964. For power tools we worked exactly as Nasu had: using only a hand-held circular saw and a hand-held power plane. With these tools we rough-cut and milled our planking to shape before final finishing by hand.

Nasu has a large collection of hand tools, particularly saws and planes. We did all the final finishing of surfaces with hand planes. He uses an axe for rough shaping and for chopping out plugs. He uses a single chisel to cut the mortises for the nails. Nasu called these mortises *daki,* a term I had never heard before.[5] Another critical tool are his *kasugai* (metal staples, or dogs), of which he had several dozen. We used these instead of modern clamps to pull parts together, align planks, and to secure boards tightly while we glued and nailed them. To hold the bottom in place we weighted it with large stones and braced it with posts against the ceiling of the workshop. When planking the sides we depended on many more props to hold planks at the proper angles. Nasu distinguished between the different types of props, calling the small props we placed under the boat *tsuku.* All other props he called either *hari* or *tsuppari,* though he most often used the former term.[6]

Nasu's most interesting tool is a special chisel for drilling nail holes called a *momigiri* or, more commonly, *moji* (see p. 34 picture). He was surprised when I told him I had never seen another one in Japan, where boatbuilders typically use a special chisel called a *tsubanomi* to cut holes for boat nails. Part of this difference can be attributed to the nails (described later) because ubune are built with nails that have a square shank, and the moji cuts a round, tapered hole. In the rest of Japan a boat nail is made from flat steel and the tsubanomi cuts a rectangular hole for these fastenings. Nasu called the more common tsubanomi a *katanomi.* Mr. Tetsuro Matsui, a researcher from Tokyo who documented Nasu's nails and moji, told me he believes there are only a few other craftspeople in Japan using this tool.[7]

The moji is exactly the same as a tool used traditionally by boatbuilders in China and is referenced in one of the classic texts on Chinese shipbuilding, *The Junks and Sampans of the Yangtse,* by G.R.G. Worcester (Naval Institute Press, 1971). It consists of a square-shanked chisel, very slightly curved, with a pyramid-shaped point. Opposite the point is a head, which is struck with a wooden mallet, just below which is a hole for inserting a wooden handle. To drill a hole the user strikes the head with the mallet, then twists the chisel, swinging the handle through an arc of 180 degrees about four or five times, then hits again.

Design and Measurement

The standard method used by researchers to measure boats today is based on modern principles of naval architecture first developed in the West. This method involves dividing a hull into evenly spaced sections from bow to stern and recording the cross sectional dimensions at each section. Dimensions are taken from either a baseline in the profile view or a centerline in the plan view. These dimensions work like an XY axis, locating points on the hull, and are recorded in a table of offsets. From this table a naval architect or boat builder can recreate the lines defining the shape of the boat. In Japan, boatbuilders have various names for the cross sections where they record these dimensions. Sections are usually called *toubun* or *yokozumi,* and in southern Hokkaido *kanaba.*

In the West, the builder usually makes temporary molds, like bulkheads, which are set on the keel, and then the planks are bent around these molds. Once the hull is planked, the molds are removed and can be used again to build the same type of boat. Japanese boatbuilders do not use molds, instead holding the planks in place with props, checking the shape of the hull at stations—usually just three—defined by cross sectional measurements and the known angles of the planks.

Unlike in the West, the spacing of these stations is usually not even. The locations are particular to each boatbuilder, placed where the craftsperson thinks they are needed to best define the shape of the hull. Typically the bow and stern sections are the same distance aft of the stem and forward of the transom, respectively, with the middle station located at the widest point of the hull. In addition, cross-sectional dimensions and plank angles give boatbuilders the information they need to lay out the lines for the bottom of the boat, along with the angles and widths of the bottom and side planks.

Nasu divides the overall length of the bottom by four, a process he calls *yotsuwari,* establishing three stations, equally spaced along the bottom, for his layout. He uses no drawings but he had a chart he had written with the plank widths

[5] Normally a mortise is a *hozoana*. Nasu contributed many unique terms to my glossary of Japanese boatbuilding. Rural regions were historically very isolated so names of boat parts, tools, and techniques are often unique to one locale.

[6] Hari is the standard Japanese term for a beam in architecture.

[7] Temple carpenters use square shank nails in roof framing and also sometimes use tsubanomi.

TRADITIONAL JAPANESE MEASUREMENTS

Like most traditional woodworkers in Japan, Nasu measures his boats using the *shaku* system and throughout this book I will use the same system in both text and drawings. There are actually two shaku scales: *kanejaku* and the slightly longer *kujirajaku*. Kanejaku is used by carpenters and boatbuilders while kujirajaku is used by kimono makers. Shaku is decimal based, just like the metric system, with 1 shaku equaling 10 *sun*, 1 sun equaling 10 *bu*, and 1 bu equaling 10 *rin* (though woodworkers generally use 5 rin as the smallest reasonable length). Tape measures and squares are readily available in shaku. One shaku is about a millimeter shorter than a foot, so as an American I have not found it difficult at all to switch from feet and inches to shaku.

Readers can keep calculators handy if they wish to convert dimensions to Imperial or metric. Relative sizes for kanejaku are as follows:

1 shaku = 303 mm	1 shaku = 11 and 15/16 inches
1 sun = 30.3 mm	1 sun = 1 and 3/16 inches
1 bu = 3.03 mm	1 bu = 1/8 inch

Two larger dimensions are the *ken*, equivalent to six shaku and *jo*, or ten shaku.

and angles at each station. He also had an interesting type of pattern I had never seen before which consisted of a slim wooden bar with three small pieces of wood mortised into it at angles. He called it his *shichizuma no kaikata*. He said he thought the term was used only by him and his father. We used the pattern to check the bevel along the edge of the bottom where the first plank lies.

Nasu said he measured an older ubune in Oze in developing his own design. It was a boat built by Ando that passed through owners in Gifu and Seki. He also repaired ubune, and some of the smaller designs he and his father built for fishermen share many of the characteristics of ubune.

Wood and Nails

Most Japanese boatbuilders use wood from a single log. The material is used symmetrically from the log, with the bottom coming from the planks sawn from the center, the next planks coming from those either side of the center, and so on. One of my former teachers explained when we were bending planks, given how thick they were, it was important they bend with the same radius to produce a symmetrical hull. Using planks symmetrically from the log, each plank was likely to have a similar grain structure, and theoretically they would bend to the same curve. As a general rule boatbuilders orient all the hull planking so the heart side of

Nasu's shichizuma no kaikata, an object both ingenious and mysterious. He made one for each specific type of boat he built. Nasu and his first apprentice are the only boatbuilders I have met who make patterns like this. To the author they represent a compelling argument for conducting research working directly with craftspeople, for they are the only people who can interpret tools like these.

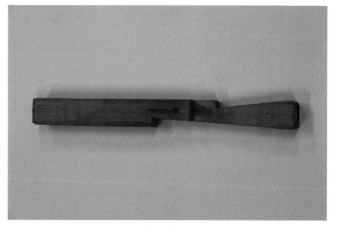

An interesting mystery: a pattern used by the late boatbuilder Ando of Gifu City, from the collection of the Gifu City Museum of History. No one—not even Nasu— knew what this was used for, but patterns are commonly made by craftspeople who have no need to label them and may in fact wish to keep their purposes a secret. One of my former teachers told me he would burn his patterns after completing a boat.

the plank faces the outside of the boat. They try to avoid or minimize the amount of sapwood in the boat, because it is more prone to rot than heartwood. Some boatbuilders have told me all wood in the hull should have the root end of the plank toward the bow, believing the root end of the tree was stronger and better able to absorb the shock of striking waves. Nasu was not familiar with this last tradition, and as you will read, his use of material was at odds with most boatbuilders.

The ubune is built out of *maki* (Japanese umbrella pine, *Sciadopitys verticillata*), a tree commonly found in Gifu and Wakayama Prefectures. The wood for our boat came from Nasu's inventory and the pieces had been stored for a long time. In fact, Nasu told us he normally stored his wood for three years before using it. He keeps his wood under cover, planks carefully separated with stickers so they dry slowly. Like all boatbuilders he shuns kiln-dried wood. His three-year wait is unusually long compared to other boatbuilders; the most common rule-of-thumb in Japan is to use wood after one year of air-drying. Nasu also insisted throughout the project we keep the boat and our materials carefully covered and protected from rain.

Maki is very light colored, slightly fragrant, dense, and very easy to plane. It is also expensive. It is commonly called *koyamaki* and is considered one of Japan's five sacred trees. Koyamaki grows very slowly. In recent years the price has risen dramatically due to a huge increase in demand from China, where the wood is used for coffins. The Kiso and Nagara River regions seem to be the only places in Japan where boatbuilders use this wood, as I have never seen or heard of it used elsewhere. Sea boats are universally planked with *sugi* (cedar, *Cryptomeria japonica*) throughout Japan, and all other river boats I have seen, from Kyushu to Tohoku, with cedar as well.

Nasu told us years ago he and his father would buy logs cut to lengths of four, five, and six meters, and bring them to a sawmill in Mino for milling. Unlike cedar, which is available in very large sizes, Maki logs were typically just 1 to 1.5 shaku in diameter.[8] Everything was flat-sawn in the sawmill to a thickness of 1 sun (.1 shaku). Nasu initially stacked his lumber vertically at his home, boards leaning in alternately from one side and the other. He called this *gassho*-style, referring to the shape made by hands held in prayer.

Nasu remembered paying 100,000 yen per cubic meter for logs, with the milling and transportation of the lumber typically costing twice again that amount. Hideki Kawajiri, Vice President of Gifu Academy of Forest Science and Culture in Mino, wrote an article on koyamaki in 2014 which reported a five-meter log suitable for boatbuilding was up to 200,000 yen per cubic meter. We used forty-nine boards for our project with a calculated volume of 1.6 cubic meters. Therefore at current market rates with milling and transportation, wood costs for an ubune would be about one million yen.[9]

For the project we purchased fifty planks from Nasu, all between four and five meters long. The breakdown for the various parts of the boat was as follows:

Bottom, nineteen planks about 3.3 sun wide
Side planking, eighteen planks up to 1 shaku wide
Sheer plank, eight planks
Plus additional thicker planks for the transoms

The Mino area is also famous for papermaking, and Nasu has built a number of wooden tubs for papermakers, also out of koyamaki. He said regular wooden bathtubs were made of koyamaki locally and he had built about twenty. Only large barrels were made of cedar. He said he also sometimes built smaller boats using cedar, as well as from a type of slow-growing cypress called *sawara* (*Chamaecyparis pisifera*) found in the Kiso and Hida regions.

Our first day on the job Nasu sorted our planking material, a process he called *itazoroe*. The planks with the most knots he chose for the bottom, reserving the clearer wood for the side planking. This might seem counterintuitive given the obvious need to keep the bottom watertight—and we would find out later how problematic these knots would be in the bottom planking—but his choice was in part aesthetic: he wanted the most visible parts of the boat made of the most beautiful material. My other teachers often made choices based the same principle, stressing that the best workmanship looks beautiful. But he also said hard knots made for a stronger bottom. Ubune often navigate

---

8   Boatbuilders who work in cedar have access to planking easily over two feet wide, so most wooden boats are planked with very wide material. The narrow planking we used for the ubune is a rarity. However, when river boat builders do use cedar, on white water rivers they prefer to plank with narrow material because this facilitates repairs in an environment where the bottom is often damaged by rocks. It is easier to replace narrower planks.

9   Or approximately $9,350.00 based on the exchange rate in April, 2020.

Nasu choosing material for our bottom planking. He chose from our narrower planks and favored material that had more knots. On the bottom these would be less visible and he believed knotty wood had more abrasion resistance.

shallow waters when the river drops seasonally, and the bottoms of these boat take a lot of punishment grounding on the cobblestone riverbeds.

Nasu did not use the material symmetrically from the log. His materials came from general inventory he had accumulated over the years. He was, however, very careful to avoid using any sapwood. Given the small logs he was buying, a typical ubune would contain wood from several trees.

Most Japanese boats have little internal framing; a strong hull is attained through the use of thicker planking. With little or no framing to attach the planking to, boatbuilders rely instead on edge-fastening the planks to each other, and the craft has developed a variety of specialized tools, nails, and techniques for this purpose.

Korean boatbuilders historically fastened their planking with wooden pegs inserted from the outside of one plank to the inside of the adjacent plank, at a steep angle through the plank seam. Edge-nailing can also be found in China. It is certainly plausible that Japanese boatbuilders were exposed to these techniques and then refined them in significant ways. Mr. Osamu Monden, a Japanese filmmaker who has researched boatbuilding and fishing traditions throughout Asia, Polynesia, and the Indian Ocean, has seen boats edge-nailed in places as far-flung as India and Madagascar.[10]

One of the distinguishing features of ubune is a unique type of square-shank nail rarely seen in Japanese boatbuilding. Three types of nails are used to build these boats: long and short *kakukugi*, .35 and .5 shaku long respectively, used to edge-nail all the planking in the hull. Nasu also called these *bouzukugi, bouzu* meaning "bald" and referring to the small head. The shorter of the two he called *harakugi,* and using an old measuring system said harakugi weigh about eight *monme* (1 monme equals 3.75 grams). The larger he called *shikikugi,* which weigh ten monme. The final type of nail is the *kasakugi*, .4 shaku long, weighing twelve monme.[11] Kasakugi are distinguished by their heads, which are set to one side of the shank at about a 135-degree angle. They are used where the sides join the bottom and in the transoms, and the angle of the head roughly matches the angle of the face of the planking when fastened. Kasakugi must be driven using a cold chisel set in line with the shank of the nail, otherwise striking the angled head one will bend the nail. A cold chisel cuts slightly into the nail, giving it a grip on the angled head. All the nails have a square shank, completely different from typical Japanese boat nails which are made of flat steel stock. The head of kakukugi are very ill-defined (the fastening is more like a large pin than a true nail), but one can see how the steel has been folded over slightly to one

---

[10]   Mr. Monden's video productions can be explored though his Studio Kobo website: www.umikoubou.co.jp

[11]   *Kugi* means nail. The individual types are named referencing their appearance: *Kaku* means square, referring to the nail head. *Bouzu* is a shaved head, again referencing the head of the nail, and *kasa* is an umbrella. Monme is an old measuring system: 1 monme is 3.75 grams or .132 ounces.

side. The head is meant to face the inside of the planking and not the outside of the boat.

Nasu told us he had about 1,000 kakukugi nails in stock but no kasakugi, so we ordered a supply from a blacksmith in Kakamigahara, Seki City. The blacksmith had never made boat nails before but had no trouble fashioning them for us, though Nasu pointed out he did not like sharp corners on the shaft of the nails. He felt sharp corners cut into the wood and risked splitting the planking. According to Nasu, the late boatbuilder Ando used to hammer the shanks of his nails to round and smooth the corners, emphasizing that he considered this an important innovation by this craftsman.

Like all Japanese boat nails ours were made of high carbon soft steel. Nasu had an interesting habit of pre-soaking his nails in brine, intentionally rusting them. He felt the rough surface created by the rust would give the nails better holding power. I also saw this done by the last boat-builder on Lake Biwa. Operating in fresh water, the nails in ubune would rust very, very slowly, easily lasting the life of the boat, so rusting them was not problematic. In fact during an earlier visit to see Nasu he told me nails had become so difficult to get he had resorted to recycling nails from aban-doned boats.

Fitting and Joining Techniques

SURITSUKE Because most Japanese boats are built without caulking, among the most important techniques of the craft are those creating a watertight fit between the planks. Fore-most among these techniques is one variously called *suria-wase*, or *aibazuri*. Suriawase is a compound of the verbs *suru*, "to rub" and *awaseru*, "to put together." Nasu used a term I had never heard before: *suritsuke*. *Noko* means "saw" and the boatbuilders refer to these specialized saws as *surinoko* and *tooshinoko* (*toosu* means "to pierce").

The basic principle of the technique is familiar to most woodworkers. To create a tight fit between two pieces of wood a saw is run through the joint. The saw cuts a parallel-sided channel, or kerf. After each pass of the saw, the two pieces of wood are moved closer together. With each succes-sive pass the high points where the planks meet are cut away and the joint becomes a progressively better fit.

Our new kasakugi, just partially rusted from soaking in brine. These were made for us by a blacksmith in Seki.

Among Japanese boatbuilders, however, the technique is highly refined, employed regularly, and among the most time-consuming and important techniques for the appren-tice to master. Like other boatbuilders, Nasu used his saws to fit the boards making up the bottom, to fit the plank seams on the sides, and to fit the planks to the forward and aft transoms. In short, every seam requiring a watertight fit is prepared by making multiple passes with saws.

Nasu insisted all sawing be done from the outside of the boat. I had never encountered this rule before as my other teachers would saw from both inside and outside. All the boat's nails were on the outside as well, and also the iron dogs (called *kasugai*)[12] we used to clamp parts together when saw-fitting were used mostly on the outside. Obvi-ously the builder has to carefully work around the dogs when sawing. We generally spaced them about six to eight sun apart when doing this work. If the gap is too loose for the saw to cut, the dog can be tapped carefully to help close the seam. If the gap between the planks is too tight for the saw, one can open up the seam by inserting a wedge. Nasu made his own wedges out of bamboo. Using a material like wood or bamboo is handy because it won't damage the saw if they come into contact. At the Gifu City Shipyard, which builds the spectator boats for ukai viewing, the boatbuilders use soft aluminum wedges for the same reason.

As the fit of the planks improves, the technique changes from a conventional sawing motion—in which the teeth are pulled back and forth through the seam—to one in which the saw is moved parallel to the face of the planks. Holding

[12] Dog is an interesting and perhaps confusing name for these clamps, which are nothing more than large iron staples. The legs splay outward slightly which makes them pull adjacent planks together as they are driven into the wood. In the West today mainly loggers and log home builders use very large dogs to clamp round material. However in Japan they are commonly used by boatbuilders in lieu of modern clamps.

The author was surprised at how much larger and coarser his boatbuilding saw was (above) compared to the fine-cutting saw Nasu used for suritsuke (below).

Close-up of the teeth on Nasu's saw (left) and the author's saw (right). Most Japanese boatbuilders use a variety of saws when fitting planks, beginning with a rough-cut and moving to a finish saw. Nasu's saws were equivalent to a finish saw.

the saw steady at about a forty-five degree angle to the work, the saw is moved forward and back in the seam rather than up and down through it; thus the same teeth stay in contact with the seam. Perhaps the best description of this action is closer to the definition of suru, "to rub."

Boatbuilders believe rubbing the saw parallel to the plank face is essential because the teeth of the saw leave tiny grooves in the wood. Grooves left by a normal sawing motion cross the plank edges diagonally from one face to the other (what will be the inside and outside of the hull), and thus can channel water through the seam. For a proper fit, the saw teeth must move in line with the plank edges so the grooves remain parallel to the plank faces.

I had brought to Gifu several of my Japanese boatbuilding saws. Nasu was surprised to see how large the teeth were.

In my previous work my teachers had used a series of saws in suriawase, sometimes as many as three sizes, working from a rough cut to progressively finer cut. Nasu's saws, on the other hand, had very fine teeth, no bigger than my finish saws.

As we began to fit adjacent planks Nasu spent much more time than my other teachers trying to get as close an initial fit as possible using the electric plane and then a hand plane. I was used to working with my previous teachers much faster on this initial fitting, as my larger saws would remove material quickly. Building other boats, my teachers and I would rough cut planking with a circular saw and then we would use our handsaws; we never used a plane. As we got started, I asked Nasu if I could use one of my saws and he looked at the large teeth doubtfully. Koyamaki is denser than cedar,

Using the uraboso to even out the marks left on the edge of the planking from suritsuke. The saw is angled so many teeth are in contact and then the user rubs the saw back and forth along the edge.

The author hammering a plank edge. The section that one is hammering needs to be securely supported, in this case by a beam.

and I immediately discovered how much more work it was to use my saw. I quickly learned to take the time to make the first fit as close as possible with the plane, the way Nasu did. When we did happen to have large gaps my saws came in handy, but we did our best to copy Nasu's methods. He said we could expect to spend two hours doing suritsuke fitting one 12-shaku plank, and he was generally correct. As a final step Nasu took a narrow handsaw he called an *uraboso* and laid it on its side with the blade lying along the plank edge. He then rubbed the saw lengthwise with the teeth against the edge. The idea was to smooth out the markings left by the previous saw. We also used the uraboso this way in tight spots where our regular saws would not fit. Later Nasu explained one can do uraboso on the plank edges to cup them slightly, removing material and thereby making the saw-fitting process a bit easier.

One critical detail in fitting the bottom planks: as each row of planking was added, we clamped them so the two edges had a slight opening of about 1 bu on the outside. Nasu said, "One bu, plus or minus zero." Getting the angle right required us to use props bracing the planks from the side. When the handsaw is run through the seam it cuts a small bevel on the two plank edges. The goal was to build the bottom with a very slight arc across the hull.

KIGOROSHI The next crucial fitting technique before fastening two planks is to pound both edges with a hammer. Called *kigoroshi*, literally, "wood killing," the purpose is to crush the fibers of the two mating surfaces. Done just before planks are clamped, glued and nailed, the two edges will squeeze against one another when fastened tightly. The joint will become even tighter as the pounded wood fibers rebound from the compression.

Japanese metal hammers have two faces, one flat and one convex, with the latter used in kigoroshi. The key to this technique is to evenly strike the edge of the planks, using overlapping blows of the hammer, and *never* touching the corners. This produces a slight concavity and when the planks are clamped together the corners of the edges are like unsupported peaks, and are crushed against one another. Any minute gap left in the seam at this point will be filled when the compressed wood fibers in the middle swell back.

Kigoroshi is ninety percent control and ten percent force, and Nasu was very strict about doing this correctly. He directed us to pound the edges a short section at a time, basically the same distance as the spacing of the dogs, first staying to one side of the edge, the head of the hammer slightly angled, then traveling back on the other side, and finally finishing by hammering up the center. The goal, he explained, was to create a smooth groove, and this meant the blows had to overlap by about 3 bu at least. They also, of course, had to be consistent, at the proper angle, and so on. Only one of my other teachers stressed this as much as Nasu; for most of my teachers suriawase was the primary technique for fitting planks.

Whenever Marc Bauer and I did kigoroshi, Nasu insisted we alternate our hammer blows. I could not think of any reason for doing this (it is surprisingly hard to coordinate one's work with another person, and Marc and I had to often stop and re-start), but later Imaishi told me Nasu explained

A close-up of a plank edge after kigoroshi. The hammer marks are visible, which is not ideal. The trick is to overlap the hammer blows and create a smooth depression down the length of the plank edge.

Another view of a plank edge. The square helps illustrate the slight groove left after kigoroshi. It is crucial to avoid the corners of the plank edge when hammering.

to her he needed to hear our individual hammering to know if we were doing it correctly. He said he could tell if we were striking too hard, too soft, or not consistently. Early in the project Nasu always checked our work, sometimes hammering small sections again, but his remarks to the researcher are an example of remarkable powers of observation—an ability to judge our work by sound alone—no doubt honed during his own apprenticeship observing his father and then working under his supervision.

As a final fit we also marked the location of any knots on the plank edge, transferring the location to the adjacent plank. Since the knots would not compress when hammered, they left high spots. We always made sure to hammer a slight depression on the edge of the adjacent plank where the knot would meet it.

KAMA SCARF JOINT The *kama* joint connecting planks end-to-end is one commonly seen in Japanese boatbuilding. It looks like a stair step.[13] In the bottom planking the length of the joint is 1.2 shaku. The vertical dimension at the end of the joint is .1 shaku and the vertical dimension of the inside corner is .15 shaku. Every effort is made to keep these joints staggered for strength, in other words, you don't want two kama joints next to each other in adjacent planks. The .15 shaku endgrain of the joint has a rounded shape. Nasu showed us how to use a nail set with a hammer to compress the wood into a concave shape. The edge meeting it is hammered convex to match. The .1 shaku endgrain is sawn at a very slight angle so when the strake is bent this angle keeps the plank joining it from springing out. The faces of the .1 shaku endgrain stays flat but both are also hammered

KAMA JOINT

The kama joint in the bottom, here viewed in profile, is fastened to the adjacent plank with three nails: one shorter (.35 shaku) at the narrow end and two longer (.5 shaku) nails at the middle and wider end. All the nails are countersunk and the nail at the inside corner is angled. Ideally the nails should exit the plank evenly spaced (value "x" shown).

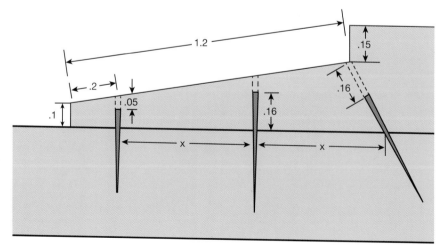

13　The *kama* is a Japanese farming tool like a sickle, somewhat like the shape of the joint.

to make a tight fit. When cutting and fitting kama joints you have to be aware that kigoroshi shortens it slightly.

Nasu also added a small detail to cutting the shape of the kama. We used a circular saw to make the long diagonal cuts for these joints but used handsaws to make the short vertical cuts. Nasu cautioned us not to cut all the way through, rather break off the wood and then chisel the inside corner. If the saw cuts even a tiny bit too far into the corner the kerf creates a hole for water to pass through the hull. He cautioned us as well about the risk of the plank splitting at the inside corner of the kama joint, something he called *tsurizake*.[14]

The kama are difficult to fit and rather than adjust the joints using a plane we would often make adjustments simply by hammering the wood. For instance, if the part you are fitting is 1 bu too long it is possible to shorten it using just the hammer. While most woodworkers would fit a joint using saws, planes, and chisels, Nasu showed us how we could use kigoroshi as well. My other teachers did the same thing, preferring to make a final fit if possible by crushing the wood, since it ultimately helped make a joint more watertight.

As we prepared to fasten our first planks Nasu asked me if I wanted to use glue or not. Boatbuilders put glue in the seams not to actually hold the boat together (the nails do that) but to keep the seams from opening up and admitting dirt. Also by filling any gaps it is an extra assurance that the seams are watertight. Many years ago boatbuilders used raw lacquer, but this was phenomenally expensive, so when waterproof glues appeared after World War II boatbuilders quickly abandoned lacquer for glue. Nasu added that glue, when hard, more or less cancelled the effects of pounding the plank edges because it prevented the fibers from swelling. Nevertheless, working as we were under very tight time constraints, I felt glue was our best insurance against leaks.

The glue Nasu uses is one I have seen used by boatbuilders throughout Japan. It is a urea-formaldehyde adhesive consisting of a white powder and an acid catalyst. The trade name for this type of glue in the West is Aerolite; I do not know the Japanese commercial name. The powder is mixed with water to form a thick paste, always in small batches because the glue has a short shelf life. Nasu always had a jar of glue ready for us and a bottle of the catalyst. Once we were ready to glue a plank we poured the amount we needed into a separate container and then stirred some catalyst in.

The top corner of a kama joint, from overhead. The slight curve is created by pounding with a round-headed hammer and a nail set. Note also the slight angle intended to trap the plank that fits into the joint. Since the side planks are all bent this angle prevents a plank from springing out.

Once the glue was spread along the edge we quickly and carefully clamped the planks together with dogs, being careful to make sure the edges were flush with each other and the slight arc in the bottom was preserved. Given the amount of catalyst Nasu added the glue would be set in less than a half-hour. Later during the construction of the bottom, in our haste we glued a plank incorrectly: it actually angled in the wrong direction. Nasu noticed it, pointed out the error, and asked me what I thought we should do. It was the end of a long day and as I looked at the plank Nasu offered me an option, "You could cut the plank off and re-do it or you could just leave it." As tired as I was my first impulse was to leave it, but something in Nasu's tone of voice made me hesitate. He is an affable and extremely friendly man but as he looked at our mistake his demeanor had changed to a deep seriousness. His question about what to do felt like a test. I said we would work late, cut the plank off and redo it. Nasu nodded and went home for the evening, leaving us to make the repair.

Near the end of the project, Nasu told me had we not fixed the plank he would have always worried about it. Hearing him say this reinforced for me what uncompromising standards he had for our craftsmanship. During the course of the project he was willing to allow slight changes in things like the width of a plank or a bevel, but the actual workmanship had to meet his standards. This subject and the importance of these principles to Nasu would come up again as we got closer to launching.

---

[14] This was a word I had never heard before, derived from *tsuri*, which means "to fish," and *sakeru*, "to break."

Dogs are used to clamp a kama joint together in various ways. Note the one in the top edge of the two planks as well as the one set at an angle, both of which pull the two planks together. For extra strength, two dogs reach over the top of the plank edge to draw it down, one placed incorrectly on the inside of the plank.

Detail showing how a nail is used to keep the plank end aligned, with two dogs holding the plank. You can see the red pencil line of the kama joint to be cut. The dogs are very close to the plank end but this doesn't matter because this section will be cut off.

Two nails angled to miss the knots visible in the bottom plank. Note how the knots near the edges are circled. We would hammer the opposing edge in order to make a space for knots on the edge to fit.

CLAMPING The iron dogs are amazingly versatile tools, and boatbuilders use them in remarkably creative ways. Nasu had dozens of them in about three or four different lengths.

As mentioned earlier, we used dogs and props exclusively to clamp planks together for gluing and nailing. Nasu liked to set the dogs at a slight angle to the surface, leaning to one side; this allowed one to swing a hammer along the plank and easily knock them loose. We generally spaced them between each nail location along the seam and, as mentioned before, always on the outside of the hull. Nasu cautioned us to not set the points too close to the plank edge for fear of splitting the wood. Typically he would alternate setting them high and low down the length of a plank seam. Near the plank scarfs it is very important not to set the point of a dog in line with the corner of the joint for fear of splitting it.

At the kama joints dogs could be angled in order to pull the joint tight and Nasu also set them in the plank edges at the kama. Here he used his longest dogs, which he called *hiki-kasugai (hiku* means "to pull"). He would pre-drill two holes by hitting one point of a dog into the plank edge and twisting it. He would make these holes just slightly farther apart than the width of the dog, so when inserted it would act to draw the two planks tightly together. They can also be set with one point over the top of a plank to draw it down.

Once two planks are clamped together we would check to make sure the inside plank edges were flush with each other. If not, one end of the dog could be hit with the hammer to both shift the plank and increase the clamping pressure. Another method for holding the end of a plank in alignment with the adjacent plank was to set a nail. Driving a temporary nail at a plank end into the previous plank kept it in alignment until the plank was fastened.

NAILING Once planks were glued and clamped and the glue set, it was time to nail them. Nasu tried to keep an even spacing between nails, generally about .6 or .7 shaku, but he made adjustments to avoid knots or we could angle nails to miss knots.

The first step in nailing is to cut the mortise. Nasu calls this mortise a *daki*, a term I had never heard before. The bottom of the mortise is sloped and at its deepest is about half the depth of the plank. Nasu exclusively uses an eight-bu (.08 shaku) wide chisel, slightly longer than most common chisels, to cut his mortises. To cut the tapered outer edges of the mortise he sets the corner of the blade against the wood and then hammers the chisel, rocking it between strikes. The chisel cuts an ever deepening groove to the base of the daki. The bevel of the chisel faces inward, and the user will discover this bevel will force the chisel to turn outward as it cuts. The trick is to let the chisel twist slightly so the sides of the mortise are cut back slightly, creating a dovetail cross-sectional shape. Otherwise the mortises have straight sides. Nasu grinds the back of his eight-bu chisels very slightly (you have to hold a straight edge against the back of the chisel to see it) which helps his chisels track in a straight line. It is a small detail but incredibly effective; it is all but impossible to do this with a normal chisel with a flat back. Once the two outer edges are cut he chisels downward in between his first two cuts, removing the remaining material. He sometimes used his regular tsubanomi like a narrow chisel to clean out the inside of the mortise.

Nasu cuts his mortises without any layout beforehand. Having cut hundreds of thousands of them in the course of his career, he can make identical mortises with extraordinary accuracy. His only measurement consists of marking the base and top. He does this by using his chisel as a measuring device. Setting the corner of the chisel at the plank seam he rolls it once, which creates a mark twice the width of the chisel (.16 shaku) from the plank edge. Setting the chisel above the new mark, he rolls it five times from this spot to mark the top of the longest mortise (6 x .08 or .48 shaku). Rolling the chisel twice, then five times he calls *ni hai* and *go hai* (literally "two times" and "five times"). Most of the planking is narrower so mortises penetrate the top edges.

Knowing we would be plugging hundreds of mortises I realized it was critical that all of us cut them consistently. Lacking the skills Nasu had developed through long experience, we either used our squares to measure and mark their shape or we used a small pattern I made of plywood, which we held against the plank and marked the shape with a pencil or knife. At first I tried cutting the sides with a small

saw but found a knife and chisel were faster. Marc purchased a chisel exactly like Nasu's and re-shaped the back of the bevel to match his. We developed our technique with the chisel quickly and marking with the knife is superior for the beginner because the kerf cut by the knife helps guide the corner of the chisel.

As mentioned earlier, the moji was a tool I was completely unfamiliar with using. Drilling the holes is difficult given the risk of driving the moji out through the face of the planking. One starts the hole using Nasu's *tsubanomi*, a chisel quite different from the one most boatbuilders use. It is important to remember the moji, tsubanomi, and the nails are all driven with a small wooden mallet, *never* with a metal hammer. The wooden mallet, which Nasu called *saizuchi*, exerts far less force on the tools and nails, and as we learned this entire process needs to be done gradually so as not to damage the tools or planking. Nasu's mallets had heads that

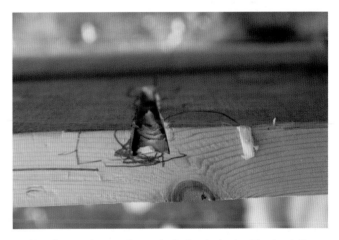

Looking down at the top of a mortise in the plank edge. The dovetail shape is clearly visible, and it is easy to understand why the plugs do not need any glue. Where planks are narrower mortises penetrate the top edge.

The side cuts of the mortise are slightly angled so the cross-section is a dovetail shape.

The author's plywood mortise pattern, clamped to the hull to guide a saw cutting the sides of the mortise. This was a short-lived experiment to try to maintain speed and accuracy, but we abandoned the idea of using a saw in favor of a knife. We did use the mortise pattern for layout, however, throughout the project.

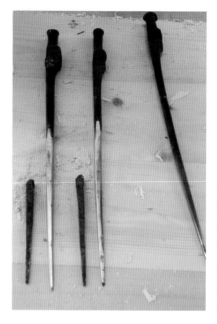

Three of Nasu's moji and two kakukugi. Note how small the heads are on these nails. Nasu's moji were slightly different sizes, which is convenient because the nails are not of a uniform length and width.

were quite small. Nasu is the only boatbuilder I have met to use a mallet in this way instead of a hammer.

Cutting a hole with the moji is a slow process. I timed Nasu as he cut a hole and set a nail: about eight minutes. While I was excited for the opportunity to use this ancient tool, given the time pressures we felt on this project I was filled with dread as I realized how slow it was. Nasu had mentioned several times that ubune have a thousand nails. By far the single most time-consuming task building these boats is nailing, and there is no way to rush this process. Twisting the moji generates an enormous amount of heat, which actually softens the metal. If one forces the tool, pounding it too hard with the mallet, the tip can break

off as it is twisted. We did this once, and other times we bent the tip of the moji as it became soft from the heat. One has to adopt a slow, steady pace (and patience) to drill holes and not damage the tool.

The shaft of the moji is tapered with a very slight curve, and it is possible to control its direction slightly as it is driven through the plank. We found starting the hole with the tsubanomi it was critical to set the primary hole very carefully, as this dictates the direction the moji will travel. One uses the moji by holding the handle at right angles to the plank face and hitting the head. Nasu typically hit the moji hard once, followed by two light taps and then, twisting the handle, swinging through at least 180 degrees,

Marc Bauer using the moji for some of the first nails in the bottom. None of us had ever used this tool before, and Nasu was surprised when I told him I thought he may be the only boatbuilder in Japan using it.

The tip of the moji is visible emerging from the side of the planking. The tool can get so hot the metal softens and bends. It has to be carefully removed and hammered straight again. The yellow marks on the moji were a guide showing us how deep to drill.

getting close to the face. I found it very useful, after driving the moji part way, to periodically put my hands on the inside and outside faces of the plank and feel for heat. If one face started to feel warm I could reverse the handle and alter the direction. To remove the moji the user strikes the underside of the wooden handle close to the metal shaft.

Of course the biggest challenge for the beginner is cutting a hole directly through the center of the plank. Nasu had a collection of moji, some in slightly different sizes, and one must choose the proper size given the nail being used. We also made a mark on the moji showing us the proper depth to drive it. The moji is driven to a depth slightly less than the length of the nail.

Earlier, Nasu had told us ideally the length of a nail as it lies in two planks should be the same proportion as the balance point of the nail. In other words, if you find the balance point, that spot should be even with the seam between the planks. In practice it is difficult to cut the holes consistently and the nails vary slightly in size as well. All of our nails drove below the bottom of the daki. With experience one can drop the nail in the hole and, seeing how far it drops in (usually about one-third to one-half way), know before hammering if it is too tight or too loose. If necessary, the nail can be removed and one can find a nail closer to the correct size, or drill deeper with the moji.

four or five times, then repeating. Twisting cuts and opens the hole so the next mallet strikes can drive the moji a little bit deeper. One will notice the moji actually backs out of the hole slightly as it is twisted.

If the angle of the moji seems incorrect from the outset, one should remove it, re-set the hole with the tsubanomi again and start over. As the hole is drilled if one thinks the moji might be headed out of the plank you can spin the moji 180 degrees and insert the handle from the other side. Reversing the curve of the chisel this way sometimes helps correct its course. One of the benefits of the tool getting so hot is one can feel the heat through the plank if the moji is

All edge-nails are inserted with the heads facing the inside. The nails, being hand-forged, were random enough that often we would pick through three or four nails before finding one that was the best fit. Nasu has another curious and unique tradition of licking the nails before inserting

Using one of Nasu's gauges to line up the tsubanomi. It is critical the initial nail hole cut with this tool is correct, otherwise the moji will cut in the wrong direction. It is very hard to correct its direction once you have started drilling the hole.

Nasu showed us the reason for licking our nails by inserting one in a hole then pulling it out, showing how it was covered in sawdust. He said the wet nail grips the sawdust inside the hole.

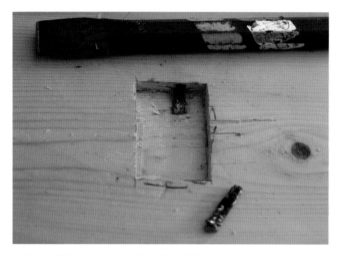

To fix a nail that came through the face of the planking we cut a mortise around it, then cut the nail tip off with the cold chisel. After pounding the shank of the nail down we fit a wooden plug to fill the mortise.

nating tradition of playing a rhythm with his mallet and nail set. A significant number of boatbuilders in Japan do this, and all say it is about slowly sinking the nail into the hole. It gives the boatbuilder more control and there is less risk of splitting the plank. The only name I've heard for this technique is *uguisu no tani watari*.[15] On an earlier visit to Nasu when he demonstrated nailing, I asked him what he called the technique and he said he had no name for it. When I mentioned this phrase he laughed and said, "Now that's a good name!"

Although only one of my previous teachers did this when nailing, my other teachers were aware of the practice but dismissed it as frivolous. Personally I think boatbuilders do this because it is fun and alleviates boredom. However, both Nasu and my previous teacher said the same thing: we were not to imitate the master's rhythm but develop one of our own. Nasu also added a nail should be driven just to the point of splitting the plank. He compared it to eating meat, saying the best tasting meat was just before it spoiled.

We did have a few nails come out through the planking. In each case Nasu put his cold chisel about five bu up from the tip and cut across the nail. Then with the nail set he pounded the tip of the nail—which folded at his cut— until it lay below the surface of the plank, deep enough so we could still finish-plane the surface of the plank without hitting it. We then covered the nail tip with ordinary oil-based window putty. In a worse case we chiseled a mortise around the nail, cut the tip of the nail off with the cold chisel, and plugged it.

PLUGS All of the nail mortises are filled with wooden plugs, called *umeki*. Nasu made them using his axe, carving them from scraps of our planking material. Just like cutting the daki, he could chop out plugs in a few seconds. None of us could come close to matching his speed and accuracy at this and, realizing we would need nearly a thousand, I asked Masashi Kutsuwa, director of the Gifu Academy furniture shop which was hosting us, to set up a jig on the tablesaw and mass produce them. In a few hours he produced several hundred plugs, but after inspecting them, Nasu shook his head, rejecting them, and went back to work with his axe.

Nasu's major role in the project was to direct us, teaching and demonstrating various techniques and also doing all

them. He said the wet nails grabbed the sawdust left in the hole and improved their grip. Noticing our hesitation, Nasu said not to worry, adding that the nails taste delicious. To generate a bit of levity in the workshop we would occasionally shout *Oishii!* (Delicious!) after licking a nail. I can reassure the reader, after having licked several hundred rusty nails, they actually have no taste whatsoever and are completely harmless.

In fact, the nails aren't driven in too hard, particularly because the wooden mallet generates far less driving force than a steel hammer. Nasu was also a practitioner of a fasci-

---

[15] Uguisu is the bush warbler, a bird famous for building more than one nest and singing frantically as it travels between them. The entire phrase means, "The bush warbler goes back and forth across the valley." The hammer's rapid rhythm is evocative of the fast-paced notes of this bird fading in and out. For video of Nasu demonstrating this technique, visit: https://www.instagram.com/p/BUw1JYAlip1/

Nasu cutting plugs with his axe. He could cut plugs this way as fast as we could install them.

the layout. Most of the day he sat watching us, supervising our work and talking to researchers and visitors. I can say, however, he personally made almost all of the plugs in our boat. He was so fast at this there were times when three of us were installing plugs and Nasu, working alone, made them fast enough to keep us supplied.

The first step is to slide the plug into the mortise and check that it is the right shape and a little larger than the tapered hole. If it drops in the mortise leaving a gap of about 2 to 3 bu at the bottom, when you hammer it in it will close the gap and fit tightly. If it is too large it will be impossible to hammer all the way in, leaving a gap at the nail head. For minor adjustments we would do kigoroshi, hammering the sides of the plug to squeeze it slightly. If necessary one can also use a chisel to carve a bit more off the sides or adjust the shape of the taper. There is no need for glue because of the dovetail shape, and with the grain of the plug at right angles to the grain of the planking the plugs will swell tight in the mortises when wet.

A finished plug. While Nasu roughed these out with his axe we would use chisels or planes to get a final fit and give them a finished surface. Note the plug tapers both lengthwise and in cross-section. They are installed without glue.

The nearly completed ubune rolled on its side for planing and finishing.

Building the Ubune

Bottom

Nasu called the bottom of our boat the *shiki,* a term used throughout Japan. As mentioned earlier, at the outset of the project Nasu was very doubtful we could build the boat in the two months we had scheduled. As we began he told us of one important marker: he thought the bottom represented one-third of the total work. He carefully picked through our material, selecting planks and trying to avoid loose knots and other imperfections. Most of our material was oriented so the heartwood of the plank faced the outside of the boat, but if for some reason reversing the plank worked better Nasu did not hesitate to do so. Our bottom was comprised of seven strakes (see drawing next page), most made up of three planks joined end-to-end. The widths of our planks varied from .42 shaku to .54 shaku.

To clarify: in the text I will refer to planks and strakes. A plank will refer to an individual piece of lumber. A full lengthwise run of planking on the boat will represent a strake. In the ubune most of the bottom strakes and all side strakes are composed of three planks joined end-to-end.

In all my previous experience in Japan, assembling the bottom strakes from the planks was done by laying them on low sawhorses on the shop floor, fitting and assembling them laid flat. Nasu does the opposite. He builds the bottom clamped on its edge, with all fitting, nailing, and kama joining to form strakes done vertically, in the sequence shown in the drawing. The center strake and one adjacent to it are done first, with nails driven from both strakes into the other. In all subsequent strakes, however, the nails come from the outermost strake into the previous strake. When the bottom is completely assembled is it laid flat, marked, and cut to shape. All nails are on the outside of the hull.

Bottom Layout

The layout of the final shape of the bottom begins by striking a centerline down its length. Again Nasu used a term I had never heard before, calling the centerline *shinzumi.* At the bow of the boat, which Nasu called *hemoto* (most boatbuilders call this *omote*) he drew an isosceles triangle, which marks where the forward peaked transom, starting at the point of the triangle, lies. From this point he measured back along the centerline. The long leg of his square measured 1.5 shaku, so he counted out this distance six times (9 shaku total) and made a mark, from which he measured back .12 shaku. This point marked the location of the first station. From the first station point he did the same thing twice more, measuring aft nine shaku with his square and deducting .12 shaku from that point. In the end this gave us an even spacing from bow to stern for stations one, two, and three: each 8.88 shaku apart. Nasu builds an ubune bottom 35.5 shaku (5 ken, 5 shaku, 5 sun) in length.[16]

At each station Nasu drew a line across the shiki square to the centerline. Along this line he laid out the points marking the outer edge of the shiki. Nasu uses no drawings

---

[16] Rather than try to measure 8.88 shaku Nasu prefers to measure out 9.0 and mark the .12 deduction. Often boatbuilders make a wooden ruler 10 shaku long with only whole shaku marked, so the square is the only measuring tool divided into shaku/sun/bu. Measuring tapes are virtually unknown among boatbuilders.

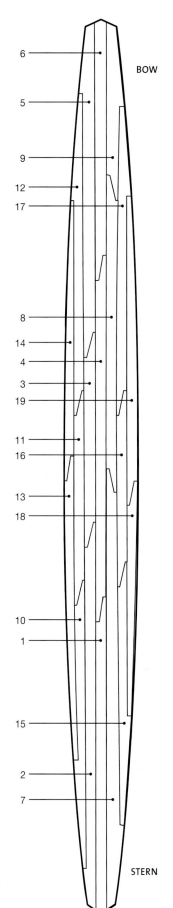

BOW

STERN

**BOTTOM STRAKE
FASTENING SEQUENCE**

The sequence of fastening
the planks together that form
the bottom (shiki). The scarf
locations are approximate.
Drawing by Migiwa Imaishi.

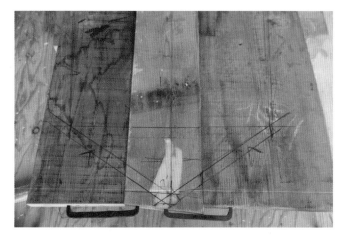

Nasu's layout at one end of the bottom. Both ends are triangles, though
slightly different sizes.

We prepare to fit the last outer planks on one edge of the bottom. This
was the first time the author had seen the bottom built vertically, a
technique which may be unique to some river boatbuilders. Note how
the center strake and the one adjacent are fastened, with nails coming
from each strake into the other. All subsequent strakes are nailed to the
previous one.

The completed bottom, just before planing.

building his boats, relying on his patterns and memorized dimensions. This is similar to many, if not most boatbuilders in Japan; each has developed unique ways of remembering dozens of dimensions. Nasu first marked a small dimension off his centerline: .35 shaku at the first station, .525 at the second, and .41 at the third. Then he laid his square and added another 1.0 shaku to this mark, which gave him the outer edge of the shiki. For whatever reason—and probably it was a tradition passed down from his father—he prefers to remember these small dimensions and then add 1 shaku. This practice may also have its origins in keeping dimensions secret, because one could write them down and simply omit the amount to be adjusted.

BEAM OF BOTTOM

Forward Station	Middle Station	Aft Station
2.70 shaku	3.05 shaku	2.82 shaku

Nasu used a batten to connect the points at the three stations. In Japan boatbuilders typically make their battens out of the same material and thickness as their planking so they will bend in a curve matching the planking material. To draw the curve from the forward and aft stations to the ends Nasu used a curious and very difficult technique of throwing

A batten held in place with dogs is used to fair the curve of the outside edge of the shiki. We traced this shape before cutting it.

SHAPE OF BOTTOM AT ENDS

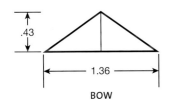

.43

1.36

BOW

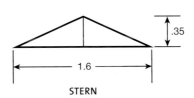

.35

1.6

STERN

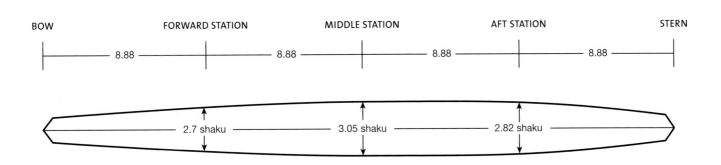

BOW	FORWARD STATION	MIDDLE STATION	AFT STATION	STERN
	8.88	8.88	8.88	8.88

2.7 shaku — 3.05 shaku — 2.82 shaku

BOTTOM STATION LAYOUT AND BREADTHS

The bottom stations are laid out equally, 8.88 shaku apart and the breadths at each station are shown, along with the dimensions of the triangles at the bow and stern where the transoms will be fastened.

The shichizuma no kaikata resting on the edge of the bottom, showing the bevel for the edge at the center. The label manaka (middle) was written by the author.

The pattern after the edge has been beveled, confirming the correct angle. This pattern is meant to always hang on the inside face of the bottom.

the string from his inkline so it struck not a straight line between two points but a curved one. This technique takes a great deal of practice because both the tension of the string and the force of the throw have to be just right, but when done correctly it draws a nice, fair curve. Many Japanese boatbuilders use this technique, called *nagezumi,* something I have never seen done in the West.

We then cut to this line using the circular saw. We made this cut at a right angle to the face of the planking but once cut it was time to plane the angle for the edge of the bottom. The first garboard planks lie against this angle, which changes slightly over its entire length. Nasu called this bevel

the *shichizuma,* another word I had never heard before. The strange patterns which had intrigued me on my first visit to Nasu's shop (see photo p. 24) are called *shichizuma no kaikata,* literally, "shichizuma shape pattern." Nasu thought it might be a term used by only him and his father. The tool consists of a wooden bar with three small pieces of wood mortised into three sides at slight angles. With the bottom resting on edge vertically one can hang the pattern on the upper edge—always on the inside face—and the small angled crossbar indicates the proper bevel for the edge. The uppermost crossbar denotes the bevel at the middle station; the next shows the bevel for the intermediate stations; and

Marc Bauer washes the bottom. This was to check for leaks through the bottom, plus it helps swell and close the holes left by the dogs.

the lowest gives the bevel at the bow (*hemoto*) and stern (*tomo*). Using the pattern we planed the proper bevel at each of the locations, then faired the bevel between them over the whole length of the bottom. The angles are derived from the hypotenuse of right triangles: the middle station a triangle with legs of .5 shaku and .2 shaku; the first and third stations .5 shaku and .175 shaku; and the region around the bow and stern .5 shaku and .12 shaku.

Next, Nasu had us lay the shiki down horizontally, with the inside face exposed, and we washed the entire surface, spreading water with a stiff brush made of twisted rice straw. When we turned the plank over we could see where water had seeped through the bottom, evidence of potential leaks in the finished boat. Most of these spots were at knots in the wood, many of which had small cracks that passed through the planking. In a few spots water seeped through the corners of our kama joints, evidence of mistakes we had made cutting them. Right before launch we would carefully go over the bottom again and putty these spots. While we looked at the seeping water, chagrined, Nasu commented that his joints never leaked.

We then rolled the bottom over and washed the outside face. We let it dry and then rough-planed the outside with just the electric plane. We turned the shiki over and did the same on the inside, then finish-planed it with hand planes.

Transoms

Normally the transom of a boat is called a *todate,* but Nasu called the two ends of our ubune *tateita.* In our case these were made of two wide pieces of maki fastened together at an angle, forming a slight peak. Nasu referred to this pointed shape as *eboshi,* since it resembles a Shinto priest's hat of the same name. It is not at all unusual for Japanese river boats to have a wide bow, unlike sea boats which have a sharp end forward. For sea boats the sharp bow is for cutting through the waves. In the case of river boats, which in Japan are often operating on swift, white-water streams, the wide bow is necessary to give the forward end of the boat more buoyancy. As the boat travels downstream a sharp bow would tend to plunge into the rapids and risk swamping. The buoyancy of a wide bow allows the front of the boat to float over the rapids.

At first glance ubune look symmetrical fore and aft. The hull measurements at bow and stern are close, but not identical, and the curve in the bottom is quite different fore and aft. In use these boats don't turn around, but most often change direction by simply backing up. If you think about it, turning a boat as long as these in a swift-moving river could

TRANSOM MATERIALS

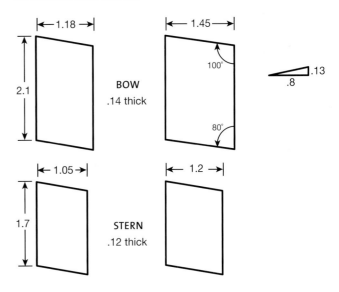

TRANSOM PATTERNS

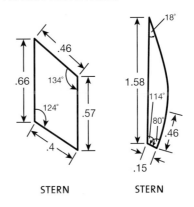

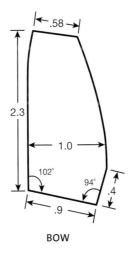

Material for the two transoms is first cut into parallelograms, with corners 80 and 100 degrees, derived from a .8 by .13 shaku triangle. One half is slightly wider than the other because the joint is a simple overlap of the two pieces.

Fitting the bevel of the transom, with the two pieces securely clamped with nails and dogs, also braced against the beam as well as a block nailed to the floor. We used this same setup when we glued the two pieces together. Note the pattern for the angle of the two halves.

After gluing we nailed the two halves of the transom together. Note the pattern which shows where the edge of the plank lines up on the other, guiding the moji. These patterns are essential guides to help the builder's nails hit their mark.

be extremely dangerous, so the double-ended canoe-like hull of ubune makes perfect sense.

The transoms were made of slightly thicker planks than the rest of the boat: .14 shaku material for the forward transom and .12 aft. We rough cut them into parallelogram shapes with the sharper corners each about 80 degrees. Nasu also brought in a very old transom to show us what he wanted. He said it may have been from a boat made in Nakaya, Gifu, during the Taisho Era (1912–26). He also had patterns that gave us the angle of the two faces of each transom when assembled as well as a pattern for the shape of the sides. The bow transom's sides meet in a peak of 134 degrees and the aft transom sides are 124 degrees. The joint in the middle is very simple: one side is beveled and the other side overlaps it, and the two pieces are glued and nailed together. This was the first place we used the kasakugi, or "umbrella nails."

Bending the Bottom

With the bottom completed, it was time to prop it solidly in place so we could begin planking the sides. Nasu depended on stout posts at two locations, each 1 shaku forward of the first and third stations. Nasu called these bracing points *tsurifuji*.[17] The props were braced between the bottom and the overhead beams of our shop. We also used weights, in the form of large stones, placed on top of the bottom planking. These main posts are set very, very tight, in fact, we had to reinforce our overhead beams to handle the load. The use of stone or concrete weights is not uncommon among river boatbuilders in Japan, but I have never seen a sea boatbuilder use them. At the Gifu City Shipyard, which builds the spectator boats for the ukai sightseeing fleet, they place large concrete blocks with an overhead crane to brace the bottom. We had watched a documentary film showing Ando building his last boat and he used many stones. Nasu said his use of fewer stones allows him to make adjustments, if necessary, to the shape of the bottom. We also put wedges under the outside edges of the bottom so the stones did not flatten the curve.

At this point we started curving the bottom lengthwise into its final shape. Nasu stretched a string across the bottom at the center station and the curvature of the bottom athwartships was .04 shaku at the middle. This was the amount of curvature we had built into the bottom. Then he stretched a pair of strings along the outer edge of the bottom starting at a point .9 shaku forward of the forward tsurifuji to a point .9 shaku aft of the aft tsurifuji. When these strings were stretched tight he wanted them to pass .025 shaku over his string stretched across the center. This indicates a very slight bend in the bottom lengthwise between the posts.

The bow and the stern of the bottom rise up considerably, however, entirely outside the posts. Nasu stretched a string from the very end of the bottom at the bow and stern, checking the height at each of the two posts. We bent the ends of the bottom up until the string was 1.15 shaku measured above the centerline at the aft post and .99 shaku at the forward post. Nasu was insistent that most of the length of the center of the bottom between posts to be almost flat, and he asked me to very carefully sight along the edge of the shiki and install short props underneath to hold the correct shape.

One final set of measurements was made to check the curve of the bottom at either end. First Nasu measured along the outer edges of the bottom from the posts to the ends, making a mark every 1.6 shaku. This yielded four marks in each corner section. He then stretched a string from the outside edge of the bottom at the aft post to the

Nasu checks the height of a string running lengthwise down the shiki (green) seeing how high it crosses over a string stretched across the bottom (red) at the center. The offset between the two strings indicates how much the center portion of the bottom is curved lengthwise.

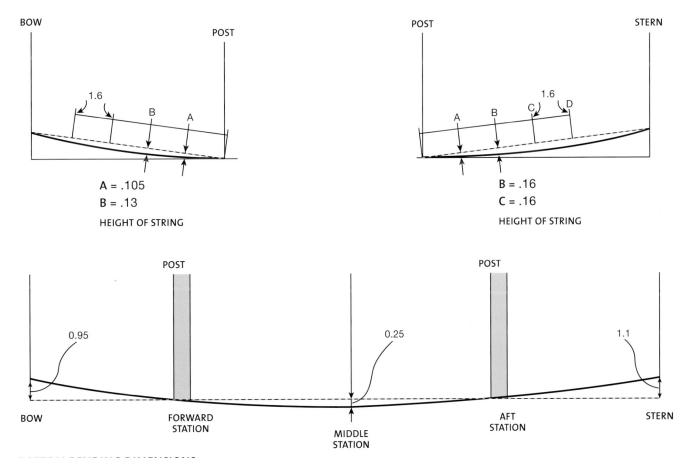

A = .105
B = .13

HEIGHT OF STRING

B = .16
C = .16

HEIGHT OF STRING

BOTTOM BENDING DIMENSIONS

A string stretched between the bases of both props allows the builder to brace up the center of the bottom until there is a .025 shaku gap in the middle between the string and bottom. At the bow and stern a string is stretched between the base of the posts and the ends with stations marked off every 1.6 shaku. At some points the builder checks the offset between the bottom and the string. Curvature of bottom is not to scale.

Nasu wanted to flatten the curve of the bottom fore and aft after we had bent the ends to the proper heights. We stretched a string across the four corners and propped the edges until we closed the gap between the string and the bottom.

Beginning to install the garboard (first side plank) on the boat. The curvature of the bottom is fixed by posts from above and below, as well as the stones.

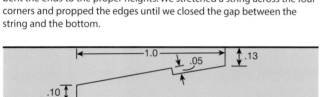

KOBERI

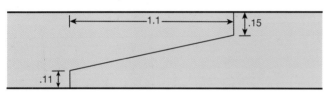

SECOND STRAKE

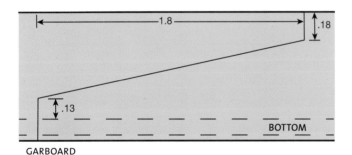

GARBOARD

KAMA JOINTS

Three types of kama joints used in the side planking.

the measurement between the bottom and the string was .105 at the mark nearest the post and .13 at the next mark.

Nasu waited patiently while I carefully looked at the shape of the bottom, making small adjustments with props. He then glanced along the edge of the bottom and made corrections. He used the term *nameraka,* meaning "smooth," to describe what the curve should look like.[18] Once again this pointed to the importance of aesthetic judgment to the craftsperson. This was also the first time I would notice how quickly Nasu was able to glance at the shape of the boat and see imperfections. This would happen throughout the project, as he could take the briefest look down the length of a plank and instantly see exactly where an adjustment was needed.

Planking the Sides

We finished propping the bottom and were ready to begin planking the sides of our boat after fourteen days of work. Nasu called the first and lowest planks along the sides the *dozuke.* The next row he called the *nimaime,* literally "second," and the next *sanmaime, or* "third." The very top planks are called the *koberi.* I had never heard Nasu's terms for the planking used before except koberi, which is widely used by boatbuilders throughout Japan.

As mentioned, in this text I refer to individual pieces of wood as planks, and a full run of planks as a strake. Using standard boatbuilding nomenclature, the first plank that fastens to the bottom I will call the garboard, followed by the second strake and third strake, using Nasu's terms.[19]

corner of the end of the bottom (the corner of the stern transom). With the string stretched tight Nasu wanted the height of the string off the bottom at the two center marks to be .16 shaku. At the bow we did basically the same thing, but used the two marks closest to the post, propping until

---

[18]   This usage is equivalent to the English "fair," how boatbuilders describe curves that are aesthetically pleasing. Boats have complicated shapes, and the experienced builder carefully takes the time to get them visually right.

[19]   In Western boatbuilding the plank after the garboard is called the first strake, followed by the second strake, and so on, until the final strake, or sheer. Here I prefer to match the strake number with their Japanese terms, and with just four planks I don't think the reader can get lost.

Using the sagefuri to set the angle of the garboard, measuring from the string on the horizontal back to a mark .5 shaku from the top of the wooden bar (mark not visible in photo).

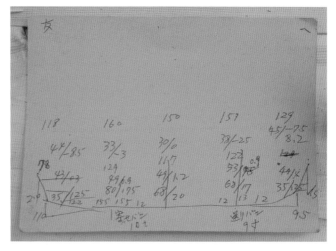

Nasu's chart of plank widths and angles (hiraki) along with other information such as the final heights of the koberi. While this represents a level of documentation, the researcher is still dependent on the boatbuilder to interpret what these numbers mean and where they are used.

Normally the top plank is called the sheer but I will use the Japanese term koberi, because on the ubune this plank is quite unique, serving also as the framing for the overhanging decks.

Like the bottom, each strake on the sides would be made up of three pieces of wood, joined with kama joints, glued and edge-nailed. On the garboard the kama were longer than those we made for the bottom, 1.8 shaku overall, and the vertical dimensions of the joint increased as well, with the upper .18 shaku and the lower .13 shaku. The latter dimension is to the top of the bottom planking; the plank extends to overlap the edge of the bottom. As on the bottom, the upper verticals are rounded and the lower are flat, but both are angled slightly to hold the middle plank in.

The side planks of the ubune rest on top of one another, a style called *nokkezukuri*.[20] The garboard angles out from the bottom, with the second strake at less of an outward angle. The third strake and sheer lean back toward the inside of the boat. Having the upper planks lean inward like this, known as tumblehome, is rarely seen in Japanese boats. Nasu brought to the workshop an offset table where he listed the width and angle at each station for all of the side planking. To set the angles of the planking we used a *sagefuri,* a tool used by boatbuilders throughout Japan. It consists of a narrow, straight piece of wood with a string with a weight attached to one end. When the wood is laid vertically against the side of a plank the string hangs straight down. There is a mark .5 shaku back from the top of the bar, and a knot in the string the same distance down from the top. The angle is recorded

A boatbuilder at the Gifu City Shipyard demonstrates an innovative version of the traditional sagefuri. The clear plastic window is marked for the various angles in the boat and the user adjusts the angle of the plank until the string aligns with the correct mark.

---

[20] This term can be translated as "on-top-of construction," what would be called carvel, or smooth planking in the West. The vast majority of Japanese boats are lapstrake construction, with strakes overlapping edges where they meet, called *wakizukuri,* literally "side construction."

Nasu's homemade tool for marking the location of nails in the edges of the shiki.

Nasu snapping a string to his plank widths, marking the top of the garboard.

as the distance between the string and the mark. Like other boatbuilders I have studied with, Nasu called this dimension *hiraki*.[21]

The garboards are held in place with dogs (again, used only on the outside) and props from both inside and outside the boat. Adjustments have to be made checking with the sagefuri to make sure the plank is at the right angle over its length. The widths of the planks are marked and the top edge is cut to a fair curve. For the garboard the bottom edge is left long, extending below the bottom, and trimmed later after the hull is completed. Right before installing the side planks the inside faces are lightly planed with the electric plane and then finish-planed by hand. The outside is left rough.

Prior to nailing the garboard we marked on it the location of any nails in the outer edge of the bottom that might interfere with our garboard fastenings. Nasu had a tiny block of wood with a carved end, which he dipped in his ink line and marked the location of the nails. We put one nail in between each of these marks. Also, these nails would be set with a slight downward angle, in order to get the heads higher. This left more wood beneath the heads of the nails, and made the garboard less likely to crack or break along the nail holes. In use, ubune often rub against rocks in the river and the garboard edges slowly wear away, another reason to keep the nail heads high. Later on in the project we would look at two historic ubune in the Gifu City Museum of History and Nasu would describe a fascinating method for repairing bottoms.

The head of the kasakugi look more like a fan than an umbrella, and making the holes with the moji we used a six-bu chisel to cut a shallow mortise for the head. This mortise is just deep enough to sink the head below the surface so we could plane the plank later. At kama joints in the garboard we angled the nails to give them greater holding power. While the side planks were being fitted we were also fitting the transoms, using planes and saws. We didn't use any glue in the garboard seam (the only place in the hull we didn't use glue or do kigoroshi), except near the final fastenings at either end. We spread glue along the last shaku at the ends of the seams and up a few inches along the edge of the transom. At this point we fastened the transoms through the garboard and clamped the transoms to the bottom with dogs. We inserted temporary nails as stops on the inside at the base of the transoms to prevent them from moving. We would not nail the transom to the bottom until the hull was finished and we could roll it on its side.

Once again, as beginners using the moji we were very concerned about missing the center of the bottom with our nails. With the lower edge of the garboard extending below the bottom it is very hard to locate where the edge is. We made a gauge to guide us, similar to one Nasu uses, but I added a feature: a hole we could use to guide a drill bit. The idea was to drill a short hole which would start our moji at precisely the correct location and angle. It made up for our lack of experience but Nasu correctly pointed out the moji doesn't remove any wood from the hole; instead it sort of grinds it up into sawdust, which he felt was essential to the

---

21   One small difference with Nasu's sagefuri is he is the only boatbuilder I have seen who marks the string (with a knot) as well as the stick. Other boatbuilders just hold their square level when measuring between the string and the mark on the stick.

Creating a gauge to guide a drill bit to start a hole at the proper height and angle in the garboard. Innovations like these were for improving the speed and accuracy of our work, something Nasu seemed to grudgingly accept. He could do all this work perfectly without any such aids.

With the gauge held tight against the bottom the hole guides the drill exactly at the right height and angle. Then we used the moji to finish the hole.

Nasu's guide is used to check that the moji is going in the right direction.

Hideyaki Goto using a saw to fit the bottom of the transom. The transoms were glued and fastened at the same time as the garboard.

holding power of the nail. The drill removes material, he explained, defeating this purpose. I argued that my invention would prevent mistakes (and speed things up, my other concern) but I agreed to drill only a shallow hole. Nasu basically shrugged and let us carry on. I am sure he understood my reasoning, but having correctly driven thousands and thousands of nails correctly he must have wondered why we couldn't do so as well. Later, when we would visit the Gifu City Shipyard we discovered the boatbuilders there also pre-drill their holes, finishing them with the moji, just like we did. They also send their drill bits to a machine shop to have them ground to a taper.

Pre-drilling our holes was a huge help to us, though we tried to drill as shallow a hole as possible. Discussing this with Nasu I asked him what sorts of innovations he had made to his father's techniques. This was the first of several times I asked this question and he said he changed nothing; he worked exactly as his father had, though eventually he did say two big changes which appeared during his father's lifetime were the use of glue and power tools. Nasu said early on his father kept accidentally cutting the cords of the power tools, and finally he refused to use them. Nasu added that working through the hot summers he used to enjoy the cool wind generated by the motor fans.

Before we had even finished nailing the garboard Nasu had picked out our materials for the second strake and began fitting the first pieces at the bow and moving aft. The process for this and all the subsequent planks is to use props to brace the plank against the outside of the hull, set the plank at the proper angle, then mark the plank widths at the stations, adjusting the height of the planks if necessary. Nasu liked to place the planks so the top edge was only 1 or 2 bu higher than the correct dimension for the plank width. With the planks propped in place he would snap lines connecting these width marks. The bottom edge of the planks are simply traced from the plank below. The plank is then removed and laid on the floor. The two edges are cut to shape with the

Nasu preparing to trace the bottom edge of the first plank of the third strake. Note how the two planks have been propped on either side of the boat overlapping the previous strake. For the top edge of the plank Nasu used an offset table with widths for the planking at each station. He would do his best to align the material so that all sapwood was removed.

electric circular saw and the fit is checked again, edges get planed if necessary, then the plank is propped in place and clamped with dogs for final fitting using hand saws.

The second strake lands on the curve of the transoms. For this reason the inside face of the planking has to be hollowed considerably to fit. We used a hollowing plane over only about .8 shaku at the ends of the planks. Nasu had an ingenious tool for twisting the planks, which he called *nejigane.* He also made the fitting easier by chiseling away at the center of the edge of the transom. By essentially carving a groove down the middle of the edge, it left much less wood to trim with the saw in order to make a tight fit. When the plank is installed this groove is covered. Nasu told me he called this method *nusumu* ("to steal"). When I told him my teacher in Urayasu did the same thing fitting frames, but called it *usotsuku* ("lying"), he laughed.

Nailing side planks to each other presents a new challenge to the beginner. First, for the sides of the boat all the nails must be vertical; none will be angled to miss a knot. I didn't realize this and cut one mortise at an angle before Nasu saw my mistake and corrected me. He said unlike the bottom the side mortises are visible and so they must be evenly spaced and uniform. Instead of angling nails to miss knots we would slightly increase or decrease the nail spacing to miss them. There was no way to fix my mistake and I tried to make light of it by saying that this lone, slanted mortise would be my signature. Nasu did not seem amused.

The ideal nail spacing for the sides is about .6 shaku in the center of the boat between the posts, narrowing to .55 shaku at the ends. It is also important not to have a nail closer than .2 shaku to a kama joint. Given these factors there may be slight variations but the overall goal is to keep an even

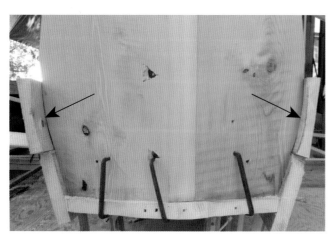

Note how the second strake is hollowed to fit the curved sides of the transom. This hollow is only planed along the final 1 shaku of the plank.

The second strake clamped in place. Note the up-and-down pattern of the dogs, and how the points avoid lining up with the corner of the scarfs so not to split them. Nails are inserted into the seam then levered downward until the inside edges are even.

The nejigane, an ingenious tool for twisting the planking.

Using Nasu's gauge to keep the moji at the correct angle. The bottom of the gauge lies flat to the lower plank, so if the angle of the moji matches it should cut directly through the center of the lower plank.

Nasu's box of lime for putting on top of the nail heads to prevent them from rusting.

The author preparing to fasten the second strake to the transom.

pattern. As we laid out the nail spacing, marking each point, sometimes we would find a problem with a knot and have to start over to make an adjustment. We used both lead pencils (gray) and red pencils and it was easy to get confused about which marks were correct.

To start our holes we held a straight guide against the face of the plank we were nailing into and tried to start our holes in line with this face. As mentioned earlier, a good check on one's work with the moji is to feel for heat on the faces of the plank. Our side nails were the shorter harakugi and the mortises were shorter as well: 4 sun. Just prior to cutting nail holes Nasu took special care to get the inside faces of the planking in alignment. This is a difficult thing to correct later so it pays to take the time before fastening to make sure the inside is smooth. Sometimes Nasu used a nail, inserted into the plank seam from the outside, to lever a plank in or out. You can also tap the outside of the plank face to shift it, or hit the top or bottom leg of a dog, which can both shift the plank and tighten the clamp. We were never to hit the inside of a plank because of the difficulty of planing any hammer marks inside the boat.

As mentioned earlier we had finish-planed the inside of the bottom before we started planking the side. Before fastening and side planks we also finished-planed the inside face. This was done because once the hull is complete it would be much more difficult to plane inside the hull.

Also, beginning with the second strake, after nailing we put some lime into the nail hole on top of the nail head. Without lime the nails would keep rusting, leaving streaks on the outside of the hull. We didn't do this on the bottom because that surface would never be seen. As he showed us this Nasu said, *omajinai,* roughly equivalent to "abraca-dabra," as if the lime worked like magic.

At the kama we put one or two nails down through the plank edge to pin the joint. These nails get set below the edge so we could do saw-fitting on the next seam. Nasu called these fastenings *momitoshi*. We also put lime on their heads and we plugged the holes with wood.

The next plank, the third strake, we began aft, opposite of the second strake. Nasu said this was because of the material he selected from our pile of planks. When choosing material, he said it was very important to keep all the kama joints separated as much as possible, though he wanted kama to fall in between our two posts, otherwise they were too close to the ends of the boat. He also chose material to keep the kama more or less symmetrical on either side. The joints side-to-side don't match exactly but they are close.

Nasu continued to choose our material for us but we were becoming more comfortable with measuring, cutting, fitting, gluing, and fastening the planking. At one point I was fitting and gluing planks and Marc spent most of a day doing nothing but following me cutting mortises, drilling holes, and nailing. He was able to do twenty-six nails. Pleased, we told Nasu this, who responded, saying in his younger days he could do fifty, although his work day might have been twelve or more hours.

Nasu began to remind us of important considerations, all driven by the fact our work would be visible and therefore had to be as nice looking as possible. He was particularly adamant about how we used the dogs. As we clamped planks to mark their shape and then later clamped them again for gluing, it was important to try, if possible, to place the dogs back in the same holes we used before. This was to minimize the number of marks left on the hull. He repeated that we should set the kasugai at an angle, leaning over. This meant they penetrated the hull less but also they could be removed more easily by tapping them with a hammer from one side. If you support the dog with your free hand while hitting they come straight out. You don't want to wiggle them from side-to-side to loosen them because it just enlarges the holes. Nasu also wanted us to set dogs using a high-low pattern, and always keep the points more than 1 sun away from a plank edge. He said we should hammer them hitting alternate corners otherwise one end tends to bounce out. Again, we were not to set the point of the dogs close to or in line with the corner of the kama joint, lest we risk splitting it. Finally, we were always to set dogs between nail locations. Since were working very quickly to clamp planks once we had spread glue on them, it was easy to forget one or more of these instructions. Yet no detail of our work escaped Nasu's gaze and he was quick to correct us.

Nasu again stressed how important it was to keep the inside seams flush. He was adept at hitting the top or bottom of a dog in order to adjust a plank. He added that as we chisel the mortise it can move the plank so he told us to keep checking the plank edges when making holes with the moji, because obviously once a hole has been made there is no way to correct the alignment.

The third strakes are the last planks before the top strakes, called *koberi,* are installed. The top plank in a Western boat

The hull after the nailing of the third strake, waiting for the last planks that make up the koberi.

is called the sheer, but given how this plank also acts to support the overhanging decks there really isn't an accurate English equivalent for the name, so as stated earlier I will use koberi. The kama joint for the third strake is shorter than those of the preceding planks, just 1.1 shaku long. The three planks—garboard, second strake, and third strake—are all trimmed flush at the two transoms but the koberi extend several shaku beyond both transoms to support the decks. At the transom Nasu pointed out its best to do the saw-fitting moving upward only. If one saws downward the saw tends to cut into the transom and there is no way to correct this. One has to carefully insert the tip of the saw into a gap. We often used the narrow uraboso saw for this. Also, Nasu asked me to trim the ends of the planking leaving 3 bu extending past the outside face of the transoms. I then planed the endgrain of the planking smooth and planed a small chamfer on the inside corners. Luckily Masashi Kutsuwa had a selection of small compass planes which came in very handy both for chamfering these inside edges as well as planing the edges of the planking flush inside the hull.

Fitting the Koberi

For the koberi the kama joint has a more complicated shape, with a "hook" in the diagonal section. The shape is reflected in the name of this joint: *hikkakegama,* from the verb *hikka-keru* ("to hook" or "to catch"). These kama are the shortest in the boat, just 1 shaku long, but they are the most complicated and hardest to cut. The shortest section of the koberi

ends are at the bow. It is not connected to the koberi with a complicated kama joint, for reasons explained below.

Nasu said this construction was exactly how his father built boats, and he went on to explain the forward deck

To create the "hook" of koberi kama, we made a series of cuts to the depth of the joint.

After cutting (left), we used a wide chisel to shape the final section of the joint.

Fitting the kama. Note the upper joint is too long in the middle. By very carefully trimming here, sometimes using just kigoroshi, one can get a tight final fit.

The initial fitting of the koberi, which extend past the transoms to support the decks.

Nasu stretching a string forward to begin the layout of the deck support. The string is connected to the top edge of the koberi at the post, and Nasu knows a height where the string passes above the transom. The end of the string (in his hand) represents the end of the koberi.

The joint in the koberi at the bow. Nasu said his father made this joint simpler than the other kama so this part would be easier to replace if damaged. The practice pre-dated Nasu's father, as the older boats in the Gifu City Museum show the same construction.

The koberi aft clamped and propped in place.

The after end of the koberi fastened. Nasu waited until the decks were installed before cutting the final shape of the koberi. Leaving this extra material made it stiffer and stronger while the decks were drilled and nailed.

supports the iron brazier used to attract the fish at night. The basket and arm carrying it are very heavy, and they are set in a hole in the deck, like a davit. The base that receives the davit is wood. Fishermen swing the basket from one side to the other when fishing, putting a great deal of strain on the deck. Nasu said it is not uncommon for this deck to break in use, so the simpler joint and shorter deck support (the end of the koberi) make replacing these parts easier.

Describing the deck structure prompted Nasu to tell us a bit about repairing boats. He said a split in the planking could sometimes be repaired by driving a dog spanning the crack and hopefully closing it tight, and then extra nails could be inserted. Of course there was a limit, he pointed out, because too many nails crowding the planking was likely to split the plank. Otherwise, he said, replacing an entire plank was a long, difficult process: cutting all the nails, removing the plank and remains of the nails, shaping a new plank to fit, and installing and refastening. Nasu did say doing repair work helped establish his reputation with customers, who in turn told others, and he felt this work eventually resulted in orders for new boats.

At either end of the boat the koberi sweep upward dramatically. Nasu laid out these curves by stretching a string from the top of the koberi adjacent the nearest post, to the very end of the plank, clamped in place. He then moved the plank end up or down, measuring the distance between the string and the top edge of the koberi at the transom. Nasu's layout of the koberi aft is as follows: we clamped a plank which ran from the post and extended 3.1 shaku past the inside face of the transom. The width of the koberi at the post was .3 shaku and at the transom it flares to .45 shaku. We attached a string to the top of the koberi at the post and

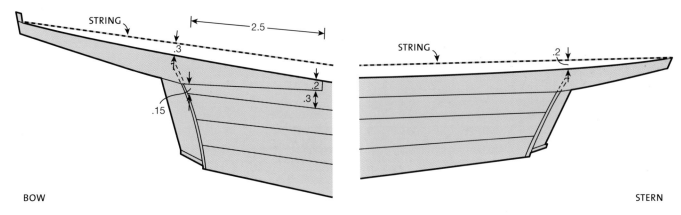

BOW   STERN

CURVE OF KOBERI FORE AND AFT

The top edge of the curve was derived by stretching a string from the ends of the koberi, bow and stern, back to the plank edge at a point adjacent to the nearest prop (tsurifuji), and measuring the gap between the string and the top edge of the koberi at the transom. At the bow this section of the koberi is short, with a simple joint, shown, so it can be easily replaced if damaged.

ran it to the top end of the plank aft. We aligned the plank so the distance between the string above the top edge of the plank at the transom was .2 shaku. At the bow we set the string up the same way and here the distance between the string and the top of the plank at the transom was slightly more, .3 shaku.

Plank Widths and Angles

As mentioned earlier Nasu brought a chart to our workshop with the plank widths at our stations, along with the values for the plank angles. Plank widths are measured on the inside of the planking and the angles are the horizontal dimensions from the string to the mark on the sagefuri.

One of the very interesting things that arose as we planked the hull was how our dimensions deviated from Nasu's chart. I wrote how Western boats are built around fixed molds, and so the tradition I was originally trained in demands the boatbuilder adhere strictly to the design dimensions, since the molds represent an absolute. What I have continually found in Japan is that my teachers were willing to deviate from dimensions based on a variety of circumstances. It could be that the materials force a boatbuilder to make a plank narrower, or where an angle gets established it turns out to be a bit different than planned. I have found that while boatbuilders may have drawings or dimensions, those are more of a guide. As a result I think of Japanese boatbuilding as improvisational: the boatbuilder has a drawing or memorized dimensions, but is not a slave to them. If the situation warrants making a slight change then one makes it.

Marc Bauer recorded our actual plank widths and angles, and when we had finished the boat I compared them to Nasu's original chart. Many of the dimensions do not match, though in truth most of these differences are slight. As we set up planks on the hull for marking, Nasu would give us a value for the angles and we would prop our plank. He would look at the plank and either pronounce the angle fine or tell us to make an adjustment. I almost never saw him refer to his chart, and I assume he had memorized all the values. We followed his instructions but I think in the end his decisions were based as much on his eye as they were on holding to any fixed dimensions. Again, Nasu's sense of aesthetics and his ability to see the shape of the boat and judge it correctly by eye trumped any hard and fast measurements.

The chart below records in shaku the dimensions of the boat we built. If a value is different from Nasu's his dimension is recorded in parentheses. Again, plank widths are measured on the inside of the hull. The dimension for the angle of the second strake aft is missing but in fact angle dimensions for all the planking at the bow and stern were largely fixed by the shape of the transom. The garboard widths are also recorded on the inside to the top surface of the bottom, so the garboard is actually about .13 wider, passing the shiki and extending slightly below. Nasu's chart only provided values for the garboard, first strake, and second strake. We recorded the koberi values based on our completed boat. Negative values for the angles mean the plank is tilting inwards (tumblehome). A zero value for the angle means the plank is vertical.

DIMENSIONS OF OUR UBUNE				
Width of plank inside / Horizontal measurement for angle				
AFT	**2nd Station**	**CENTER**	**1st Station**	**FORWARD**
koberi				
(as built) .48 / −.095	.35 / −.11	.31 / −.10	.30 / −.08	.49 / −.11
third strake				
(as built) .44 / −.085	.31 / −.03	.31 / 0	.355 / −.03	.46 / −.08
(Nasu) / (−.080)	(.33) /	(.30) /	(.38 / −.025)	
second strake				
(as built) .50 / .03	.50 / .08	.52 / .09	.53 / .08	.50 / .05
(Nasu) (.43) /	(.49 / .09)	(.49 / .12)	/ (.09)	
garboard				
(as built) .35 / .12	.80 / .18	.70 / .20	.70 / .18	.35 / .125
(Nasu) / (.125)	/ (.175)	(.68) /	(.68 / .17)	

Fishermen swing the davit from one side to the other when fishing, which puts a great deal of strain on the deck structure. Nasu said it was common to have to repair or replace these parts.

Deck Layout

With the hull complete, Nasu took us all to the riverside in Seki to see the ubune there. Most were built by his former apprentice. He wanted us to see the layout of the decks, called *hiraita,* and understand the function of all the parts before we installed them on our boat. Dark green boughs were tied in bunches and floating around the boats. Nasu identified the plant as *mukuge* (Rose of Sharon), and pointed out the usho had gathered them. Before fishing they would pack these branches around the davit in the hole, where they act as a lubricant.

At this point I asked Nasu about any ceremonies or traditions practiced by boatbuilders. I mentioned a ceremony which often takes place when the bottom is finished called the *chounadate,* when the boatbuilder purifies their tools

Ubune in Gifu showing how the davit is wrapped in Rose of Sharon to lubricate the hole.

The old transom, showing a pattern for the central angle, slightly different between bow and stern, and to the right a pattern for the curvature of the aft transom's sides.

Nasu shows Migiwa Imaishi of Tobunken and Satoshi Koyama an old transom and deck we used as a model to make our own.

and prays for the successful completion of the boat and the safety of the builders. Nasu recognized the tradition, which he called a *kikoshiki,* but said only sea boat builders did this. He did remind us about the process at the outset where he chose our planking and he said some river builders would only do this on an auspicious day and this would act something like a ceremony. He pointed out this was a tradition he did not follow.

While we were finishing the koberi, Satoshi Koyama had begun building our decks, copying an old one Nasu

had brought in. The two planks forming the deck are nailed together from the bottom at a very, very slight angle so the two sides angle downward toward the center seam. These nails are set from underneath and not plugged. I started planing the material, but Nasu stopped me: he explained the bottom of the deck isn't planed smooth because it was important to keep this area rough as a good habitat for spiders. Spiders are the emissary of the sea god so it is advantageous to encourage them to inhabit the boat. Also, the owners of ubune nail one or two *ofuda* (a small wooden tablet from a Shinto shrine) to the bottom of the forward deck, preferably one from Konpira Shrine in Kagawa Prefec-

The aft deck, viewed from underneath, showing the four nails which hold it together. The hole has only been cut on one half and will be enlarged later. Also visible is the rough bottom surface and unplugged mortises, left to provide a home for spiders.

The aft deck is glued and clamped with dogs. The davit hole is visible. Later the koberi will be trimmed down close to flush to the deck.

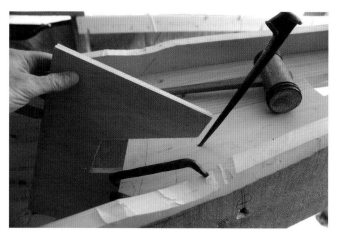

The author made a gauge similar to Nasu's in order to find the location and angle of the holes nailing the deck to the transom.

ture.[22] Nasu said the ofuda should be replaced yearly but he didn't think today's usho did this.

Fitting the two decks between the koberi is fussy work. One has to get a good fit and also fit to the transom. The latter is end grain which is difficult to plane, so I used a chisel and removed material from the center of the transom's top edge, the same nusumu method we used to fit planks to the side of the transom. Once completed the decks are glued and nailed to the planks and transom with kasakugi.

The forward deck has a .3-shaku diameter hole for the fire basket davit located just forward of the peak of the transom. In order to strengthen the deck at the hole, three small beams are installed underneath. On top two beams are fastened at angles on either side of the hole with another beam across the deck over the transom. The beams surround the hole, their inside edges positioned 5 bu away from the edge of the hole. Because the davit will rub primarily against these beams it is important for them to be made of an oily wood, like red pine. Gifu Academy's furniture shop had some material, but Nasu felt it wasn't oily enough. Mr. Yoichiro Adachi, an usho from Seki, brought us some red

Yoichiro Adachi, the usho from Seki, with the sections of red pine temple beams he donated to the project.

---

[22] Konpira Shrine, also known as Kotohira, is located on a mountaintop in Nakatado, Kagawa Prefecture. One climbs over 1,300 stone steps to reach the main shrine, considered the patron shrine for sailors.

The beams installed under the hiraita to reinforce it. Note these were fastened with modern, stainless steel screws.

The cross-beam is fitted and the other beams fit into mortises. The three beams surround the davit hole.

pine he had recycled from a temple beam, which Masashi milled in the furniture shop.[23]

Nasu had me draw an arc for the back edge of the deck, and following his instructions I cut this with a gouge and finished it with a chisel. The boats in Seki built by Tajiri, Nasu's apprentice, have a different shape: two arcs which form a peak in the middle. Nasu said this acts like a signature for Tajiri's boats, which are otherwise indistinguishable from his master's, but Nasu thought the peak would be uncomfortable because the usho sometimes lean against the deck when fishing. On the subject of whether or not he signed his boats, Nasu told us he carved the year the boat was built on the underside of the main beam. Otherwise he made no marks identifying himself as the builder.

The finished bracing.

We cut a square hole in the aft deck. Called the *ebisuana*, combining Ebisu, the god of fishing and commerce, with *ana*, meaning "hole," it is used for mooring lines. We also wrapped the end of the aft deck in copper plate. We had studied the way the boats in Seki were coppered: the metal plate was bent over on the sides first, then bent over the front, the edges folded underneath and fastened with copper tacks. I did this by first folding a sheet of paper over the deck end, and when I was satisfied, I used it as a pattern to cut and fold the copper.

The final detail on the bow transom is the installation of a small, trapezoidal piece of wood standing upright at a slight angle at its very end. This is called the *sansubo,* but Nasu was quick to point out he thought this was yet another term used only by he and his father.[24] Once fitted, two

Nasu marking the end of the aft deck.

---

[23]   Adachi has the distinction of being from the longest lineage of usho in Japan. He is the eighteenth generation of his family to fish with cormorants.

[24]   Ando Gosaku, the former builder of ubune in Gifu, also used this term according to research by Migiwa Imaishi.

Nasu trimming the bottom of the koberi at the stern.

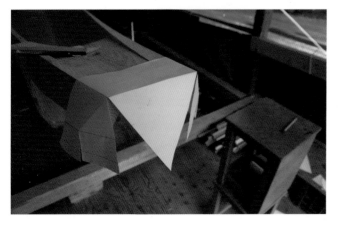

The author's paper pattern for the copper plate covering the end of the aft deck. Making a pattern out of paper helped us determine the proper shape to cut the copper.

The copper on the deck end when finished, viewed from above.

The same copper, viewed from below, showing how it wraps around the koberi.

Satoshi Koyama fitting the sansubo at the end of the bow transom.

A view of the dovetail joint from the bottom. The ariotoshi part fits into the top of the koberi. The part labeled .5 (.05 shaku) lies above the planking.

The beam installed and nailed to the koberi with kasakugi.

Copper covers the entire joint and beam end and drops down over the planking. It is important to fold the copper in such a way that rainwater will always drain off the boat and not be trapped under the copper.

The removable framework under the main beam.

kasakugi are driven down through the top of the sansubo, through the deck, and bent over where the points emerge. Nasu acknowledged this was fragile, often breaking off, and was merely ornamental.

Amidships Beam

The only permanent internal framing in the entire boat is a single beam spanning the hull near amidships. We made this beam out of *hinoki* (Japanese cypress, *Chamaecyparis obtusa*), and it connected to the hull with a joint he called *ariotoshi* but in other contexts is called *arikake*. The root word *ari* ("ant") refers to the triangular shape, like an ant's head; in English we call this a "dovetail" for the same reason. It is difficult to define this joint precisely because normally the two pieces of wood being joined are flush with one another. In our boat the beam actually remains slightly higher than the koberi. Also, the material above the koberi is not cut to the dovetail shape but left square, effectively hiding the dovetail joint. After a bit of discussion Nasu called this joint *ariotoshi no kabuse* which exactly describes it, as *kabuse* means "covered."[25] We fastened the joint with two kasakugi nailed from the top through the beam and into the edges of the koberi. Then both beam ends were wrapped with copper, covering the entire joint, including the nail heads, and extending down the outside of the planking.

Koyama also made a floor timber to rest across the bottom of the boat under the beam, supported by vertical timber. This would stiffen the bottom slightly but Nasu had him make it so it was removable. Satoshi also made six small beams notched to fit over the koberi. These help hold the shape of the hull when the boat is not in use, but are removed before fishing. The lack of internal framing in the ubune is really quite astounding, given the size of the boat, but this

---

[25]   In the parlance of Western woodworking perhaps this could be called a blind dovetail.

Koyama beginning to make the temporary beams which hold the hull's shape when the boat is not in use.

was another adaptation demanded by the customer, because the usho need as much clear space as possible to handle the birds, catch, basket, and fuel for the cresset.

At this point Imaishi walked around the boat making a careful count of the number of nails we used:

KAKUKUGI		KASAKUGI	
Bottom (shikikugi)	277	Transoms	20
Sides (harakugi)	384	Garboard	160
Decks	18		
Total	679	Total	180

Grand Total: 859

Nasu showing us how to caulk the joint at the garboard.

At this point we removed the stones from inside the hull and took down the posts

Caulking and Finishing

Makihada is the traditional caulking used by boatbuilders throughout Japan. It is driven into the seams between planks *from the inside* to make them watertight.[26] It is typically made from the inner bark of the cypress, though I was surprised to see for the first time makihada actually made from maki, the species we were building the boat with. Nasu said the cypress version is more flexible. For our boat, the only seam without any glue was between the garboard and the bottom. Working from inside the hull, first Nasu showed us how he opens a slight gap between the planks by hitting his caulking iron in the seam, providing some space for the makihada. He called his caulking iron a *hadabera* (*hada* is the caulking and *hera* means "spatula") but throughout most of Japan a caulking iron is called a *yatoko*.

Makihada looks like a loosely woven rope, and before driving it into the seam one must first peel away a small strand and roll it between one's hands, or knead a ball of it to make it soft. Earlier in the project Nasu's wife told us that throughout his career she almost never looked in her husband's workshop, but she still played an important role because she made his makihada. In fact, during our project her daughter Setsuko arranged a workshop in which her mother showed how to make makihada.

---

[26]   Caulking from the inside will no doubt surprise some readers, since Western boats are always caulked from the outside.

Traditionally in Japan new boats are not caulked. Only as boats age and begin to leak do boatbuilders caulk seams. Nasu explained that given its length and relatively thin planking, ubune are actually quite flexible, and any seams without glue can open up on the river, so he did caulk his boats when new. We drove makihada into the garboard seam and about .5 shaku up either side of the transoms. The material is quite fragile, so it is important to not hammer in one spot too much. Nasu had a very rapid cadence of quick, sharp blows with his hammer, rocking his caulking iron as he moved along the seam. It is crucial to angle the caulking iron so it is directly in line with the seam; the goal is to push the makihada between the planks and not cut into the wood with the tool.

Most of the mortises we cut to nail the koberi are visible along the top edge of the plank. Nasu stressed he wanted the visible endgrain of the plugs to be of uniform thickness when looking at the plank edge. We carefully planed the outside of the koberi to try and trim the plugs. After doing some trimming Nasu decided the thickness of the top edge of the koberi would be .0875 shaku. We set his marking gauge to that distance and made a mark to guide our final planing. He said toward the bow and stern we could leave the koberi a little thicker since the usho would be working in these area and their lines would wear away at the planking.

With the boat caulked and beam installed, we were able to roll the boat on its side and finally fasten the transoms through the bottom. We were also able to trim off the bottom edge of the garboard, a job Nasu started using his adze, mainly for the benefit of some visiting researchers who wanted to see him use this tool. When he got tired I took over, but after a few minutes I asked Nasu if I could I finish it with a circular saw and plane. He quietly said, out of earshot of the researchers, using the saw was fine and that was how he usually did it. The garboard's finished edges hang about 4 bu below the shiki, the edges beveled parallel to the face of the bottom. We also put a bevel on the inside corner only of the garboard's exposed edge.

We chiseled out several bad knots in the garboard and glued in plugs to fill them. Earlier we had splashed water over the outside of the hull to raise the grain prior to finish-planing and also to make the wood swell and close the small holes left by the dogs. When it was time to plane the entire outside of the boat, we sharpened our selection of planes and got to work, planing one side and then rolling the boat over to do the other side. While nailing the side planking we had circled with a pencil places where we thought nails might be close to the surface. At these spots we were very

Nasu's caulking iron in the seam. He had some of these tools with bent blades which was convenient for getting into the corner between the garboard and bottom.

After marking the top edge of the koberi with the marking gauge, we planed to this mark to create a consistent edge along the length of the boat.

Nasu using his adze to trim the bottom of the garboard. We have started to plane the outside of the hull.

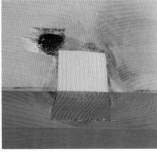

Several bad knots, like this one in the garboard, were cut out. Plugs were made to fill them then glued in place.

The putty we used for all *kasakugi* nail heads, except the garboard where we used a harder putty.

careful to remove just enough material to make it smooth. Nasu watched us, saying *kugi konnichiwa* ("Hello nail!") as a humorous warning to us while we worked. We inspected the bottom and pushed putty into any tiny cracks which potentially could leak. Nasu uses ordinary, oil-based window glazing putty; he told us in the old days he used to make his own putty by mixing lime and rapeseed oil. We used a harder type of putty to cover the nail heads in the garboard because these were below the waterline, and then used the

The exterior of the hull was planed entirely by hand.

The ubune essentially finished, except for puttying nail heads and installing copper.

window putty over the kasakugi heads elsewhere in the transoms and decks.

A final step involved cutting a small bevel on the top edge of all the plugs in the on the sides of the boat. Nasu wanted us to chisel a slight bevel so water on the hull would tend to drain off the exterior of the boat and not flow back behind the plug. It is a small detail but upon reflection very important.

At the request of the customer, we ran a strip of fiberglass tape over the outside garboard seam, tucked into the corner between the garboard and the bottom. Kentaro Hiraku, the boat's new owner, said his teacher discouraged the use of fiberglass covering the entire bottom because he felt the boat's bottom needed to be flexible for safety, and a stiff bottom couldn't absorb the shock of hitting waves. After several years if the bottom showed deterioration, applying a full layer of fiberglass cloth will still be an option to extend the life of the boat.

Nasu brought to the workshop one of his bailers to serve as a model for us. Koyama copied it while I made another one, drawing inspiration from a unique and unusual bailer I had seen in one of the boats in Seki. The scoop was the same as Nasu's normal bailer but the handle was about 4 shaku long, so the user could scoop water from the boat without bending over. I have never seen a bailer like it, but

(Above) Note the tiny bevel at the top of the umeki, cut to let water run off the hull and not behind the plug. (Right) Bailer made by the author. The scoop is the same design as Nasu's but with a longer handle.

it seemed like a brilliant idea, and much easier to use than a typical bailer.

Nasu's bailer design flares outward to the opening. The handle is mortised into the back of the bailer, and these are the first parts made and connected. Then the curved bottom is sawn and planed, with a rabbet joint along the back. The

The finished boat.

back is then put in the rabbet in the bottom and glued and nailed together. This assembly is then used to trace out the curve for the sides, which are cut, fit, and fastened.

The construction of the ubune took forty working days. Nasu had told us the bottom represented one-third of the work; we built it in fourteen days so he was almost exactly right. Nasu was with us at least five days a week advising, teaching, demonstrating, and doing all the layout. He assisted with the actual construction from time to time, among other things cutting many of the mortises and making almost all our plugs. Marc Bauer and I worked full-time, often more than eight hours a day, often on Saturdays, while Satoshi Koyama and Hideyaki Goto worked part-time. Various other people joined in to help at times including Migiwa Imaishi, Masashi Kutsuwa, and Masashi's father.

Launching

As we got closer to our launching Nasu continued to stress our workmanship had to be at its best. He said his association with our project meant his reputation would be judged by the quality of our finished product. It is worth remembering that we researchers are just visitors; we don't live in the communities where we study, but our subjects do. It is easy to think of documentation as a passive exercise in observation, but in our case, we also had an obligation to protect Nasu's reputation as a craftsperson. Although his admonitions throughout the project added to my stress over the schedule, I am grateful to Nasu for reminding me of this responsibility.

During one of our earlier interviews Nasu said launching ceremonies were always at the discretion of the customer,

Researchers, boatbuilders, and volunteers gather as the ubune is loaded on a truck for its trip to Gifu City and the launching.

The owner and builders of the ubune prepare to capsize the boat three times, in accordance with local custom, after launching in the Nagaragawa in Gifu City, July, 2017. From left to right: Kentaru Hiraku, Satoshi Koyama, the author, and Marc Bauer. Photo by Masashi Kutsuwa.

not the boatbuilder. He said some customers might hire a Shinto priest while others would conduct the ceremony themselves with just salt and sake, called *omiki* when used ceremonially. Others would do nothing at all.

Our launching took place in the center of Gifu City on July 22, 2017, in front of the Nagaragawa Ukai Museum. The museum hosted lectures and a reception party after the launching. We learned one final statistic as the boat was loaded aboard a truck for the journey to Gifu: the crane operator lifting the boat recorded a weight of 400 kilograms (882 pounds).

Elsewhere in Japan these ceremonies are Shinto-based, but always reflect some sort of local custom. They are not always conducted by a Shinto priest; just as common are ceremonies conducted by the boatbuilders themselves, with the participation of the boat owners. The general tenor of the ceremony is to purify the boat and ask the Shinto spirits (*kami*) for the prosperity and safety of the owner. Salt and sake are used to purify the boat, with fish and rice cakes (*mochi*) the most common shrine objects offered as gifts to the kami. Often fishermen have their own local customs they perform once the formal ceremony is done and the new boat is in the water.

Our ceremony in Gifu was very simple, with absolutely no Shinto-related activities. The traditions in Gifu originate entirely with the fishermen themselves, who take the boat onto the river and then capsize it three times for good luck. The local belief is a boat capsized this way at launch will never capsize again. However, according to Masaaki Kon, a researcher at Kanazawa University, other places in Japan have a tradition of capsizing boats at launch.

Nasu added a new interpretation to the local tradition of capsizing the boat. He thought fishermen may have done this as a quality check, and if the customer found mistakes they could begin to negotiate a discount from the boatbuilder.

Conclusion

Seichi Nasu's life spans an epoch between the privations of the immediate aftermath of World War II and the emergence of Japan a generation later as the world's second largest economy. It was perhaps the most unparalleled transformation of a society in modern history, and it seems almost no corner of Japan was left untouched by the forces of modernization. The Nagara River and Kiso River Valleys saw their communities change with the influx of small manufacturers and the bright lights of suburban stores. The beautiful architecture in the center of Mino City was luckily spared, but small family businesses and shopkeepers were eventually pushed out. Today the town's charming architecture attracts tourists and Mino's central businesses largely cater to them. The wooden boats on the local rivers were too specialized to become a target for mass-produced boats, but even here, near the rivers' headwaters, the culture could not escape the consequences of modernization. The impact of dams built downriver near Nagoya reverberated on the environment all the way upstream, and the number of ayu diminished rapidly. Ayu ceased to be viable commercial fishery forty to fifty years ago, and with it began a decline in the number of boats ordered. Now the river is filled with sport fishermen who park their automobiles on the shoreline and wade into the river. The era of fishermen, boatmen, and their wooden boats is largely gone.

Fishing with cormorants became another tourist industry, though at its center the usho stand as a bulwark against change. In their traditions and family lineages they try to maintain a practice well over a thousand years old. The ubune are obviously integral to this culture, and if one looks at just the usho, the cormorants and boats, the scene they create would be indistinguishable from that of the

Meiji and perhaps even the Edo eras. In the face of change they are like mathematical constants, never changing while the world changes around them. Unfortunately, the future of ubune building is tenuous at best. In the past there was always a younger craftsperson ready to take over when the elder craftsperson retired. There was also enough work in general—building a variety of fishing boats—for a boatbuilder to maintain a workshop, tools, and an inventory of materials. But now with just nine usho between Gifu and Seki, building ubune cannot come close to supporting a full-time boatbuilder.

As stated in my preface, the goal of this book was to record as faithfully as possible the construction of ubune. To anyone interested in building these boats, or any other type of traditional Japanese boat, it is obvious that a book of any length can never replace a six-year apprenticeship. Working alongside Nasu, hearing his stories, and witnessing his skills, the goal of documenting his work feels impossible next to the enormity of his experience and the subtlety of his understanding of his craft. This was the power of the apprentice system, long and arduous as it was, coupled with a lifetime of hard work. It wasn't an easy life, but it produced craftspeople of remarkable skill and intuition. Unfortunately, we no longer live in the kind of world that sustains such craftspeople. If the craft of wooden boatbuilding is to survive it will be due to committed work by researchers, museums, schools, and students.

The changes that have swept Japan we invariably call progress, and we measure it by what we have gained. Rarely do we pause to think about what we have lost. The broader changes of the Nagara River region have driven the craft of wooden boatbuilding to the brink. All of us who were involved in this project were very fortunate to have been the recipients of Nasu's generosity of spirit and his willingness to share his experience and skills with us. I believe the best way we can honor his gift is to do everything we can to preserve this craft and find some way to maintain these skills for future generations.

Satoshi Koyama takes one last look at our ubune as Kentaro Hiraku, the boat's new owner, motors away after delivery following our launching.

Illuminated by the fire an usho extracts a fish from one of his birds on the Nagara River in Gifu.

Glossary

The following boatbuilding terms were used by Nasu as we built the boat together. Those labeled with an asterisk were either completely unfamiliar to me based on my previous research or identified by Nasu and others as local (Mino dialect) terms. It is understandable that many of the name of boat parts will be unique, since ubune are a specialized watercraft. Some boat parts have no English equivalents.

NASU'S TERMS	OTHER JAPANESE TERMS	ENGLISH
akatori	akashaku/akabishaku	bailer
aritoshi *(ariotoshi no kabuse)	arikake	dovetail
bari	tsukaibou/tsuppari/tsupparibou	beam/post
bouzukugi		type of nail
bochibochi		gradually
chouna	ono/teono	adze
daki*	hozouana	mortise
dashizan*		davit supports
dennoko	marunoko	electric circular saw
dozuke*	shitadana/kajiki	garboard plank
ebisuana*		square hole in deck for lines
eboshi*		triangle shape
gomakashi*	anaume	plugging a hole
hadabera*	yatoko	caulking iron
hai/mai		counter for widths
hemoto*	omote/hesaki	bow (location in boat)
hikkakegama		hook scarf
harakugi		type of nail
hiraita*	kappa/kanpan	deck
ikihiki*	aijurushi	alignment marks
ippuku	kyukei/oyasumi	work break
itazoroe*		choosing planks
kagariana*		hole for brazier davit
kakukugi*		type of nail
kama		scarf joint
kaminoke		a hair's width
kasakugi*		type of nail ("umbrella nail")
kigoroshi	kigoroshi/korosu	wood killing
kikoshiki*	chounadate	keel-laying ceremony
kirikomi		saw kerf
koberi	uwadana	sheer (top) plank
koguchi	chiguchi	end grain
kugijirushi*		tool to mark nail locations
kusabi		wedge
makane*	chokakku	right angle
maruganna		round-bottom plane

mawashi*	jumbi	preparation
mebunryou		measure by eye
mentori	men	chamfer
momidoshi*		nails through top of kama
monme		traditional weight measure
mukuge		Rose of Sharon
nakabari	funabari	beam
nagezumi		throwing marking line
nameraka		smooth, fair
nejigane		bending tool
nimaime*		second side plank
noseude*		short plank supporting forward deck
ogakuzu	kuzu/nokokuzu	sawdust
omajinai		casting a spell
sagefuri		plumb bob angle tool
saizuchi		small wooden mallet
san		cleat, crosspiece
sanmaime*		third side plank
sansubo*		decorative block
sechakuzai	bondo	adhesive
sekkai		lime (mineral)
sendousan		boatman
shichizuma*	ne	plank bevel (on bottom)
shiki	soko	bottom of boat
shikiosai*	shikisan	floor timber
sokoshinzo*		making new bottom (repair)
suihei	mizumori	level, horizontal
suriage*		lifting the bottom (repair)
suritsuke/surigakari*	suriawase	saw-fitting
surinoko/toosunoko		boatbuilding saws
tagane		cold chisel
tateita	todate	transom
tebiki	nokogiri	hand saw
tomekugi		stop nail
tomo	tomo	aft / stern
tsukesage*		new side plank (repair)
tsuku*		vertical post
tsurifuji*	tobun/yokozumi/kanaba	station
tsurifuji*		bending point in bottom
tsurizake*		to break at a corner
ude*		arm, support
uguisu no tani watari		rhythmic nailing
ukezan*		davit support
umeki		plug
uraboso		narrow saw
ya	kusabi	wedge
yokoyama*	chokkei	diameter

The Gifu City Shipyard and Museum Boats

A few days before the launching we visited the Gifu City Shipyard, which builds and maintains the fleet of spectator boats, numbering more than forty vessels. The two largest can carry fifty passengers, while most of the boats carry between fifteen and forty passengers. The shipyard is fully equipped with power tools, an overhead crane, and modern clamps. The boatbuilders use a common Japanese glue with the trade name Bondo and nail their bottoms with the more common flat steel boatbuilding nail rather than the square-shanked kakukugi we used (although they use kasakugi the same ways we did). I was intrigued to see that to make the holes for the nails along the garboard fastening to the bottom they used an ordinary drill bit, ground so that it tapered to roughly match the shank of the kasakugi. They also had moji and tsubanomi. The boatbuilder explained they would drill first and then insert the moji and cut the hole a little bit deeper, a process that was obviously much faster. Also, the shipyard made the transoms for their boats out of steel plate, bolting them in place through the hull planking.

Seeing the drill bit and remembering how time consuming it was to use the moji, I asked one of the boatbuilders if they had ever considered abandoning the moji and kakukugi and using a drill bit and fastening with screws instead? The boatbuilder was pessimistic about this, saying nails were a better fastening for river boats, but then he admitted it had been done once building a tour boat about twenty years ago.

Gifu City Shipyard, where the spectator boats for ukai viewing are built. These are much larger than ubune, though the overall methods and techniques are similar. The photo shows the bottom for a new spectator boat.

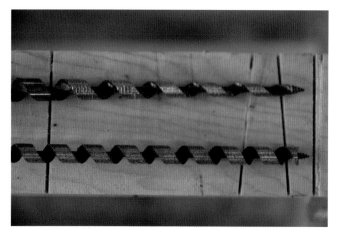

Two drill bits, the upper one ground to a taper at a machine shop, the lower one of standard design,. The shipyard uses several innovations on traditional methods to speed the construction of their boats.

I asked if the experiment had worked and he pointed out that particular tour boat was still in use.

From the shipyard we walked over to the Gifu City Museum of History where we had arranged with the curator to see two old ubune in museum storage. Both boats were in very good condition. We noticed some significant differences from our boat. First, the center plank of the shiki flared outward at the ends, which was very curious. This would seem to make planking the bottom more difficult, and I cannot think of any practical reason to do this. Also, we noticed how all the bottom seams were caulked from the inside, so these boats were built without glue. The planking

was also significantly thinner on these older boats: the sides were just .065 shaku and the bottoms .080 shaku. Except where we thinned the top of the koberi (.0875 shaku) all the planking in our boat was at least .090 shaku thick.

I found these scantlings astounding. An ubune is forty-two feet long overall, with but a single beam for framing, yet the planking is only slightly over an inch thick. From an engineering standpoint their construction seems impossibly light, and here were two historic boats built even more lightly!

Looking at the garboard of one boat, Marc noticed the remnants of chisel marks along the bottom edge. As we studied them we realized they were the remains of the mortises for the original set of nails, and the bottom of the boat had been lifted up and re-nailed, then the bottom edge of the garboard trimmed off. When we saw Nasu again we asked him about what we had seen. He told us raising the bottom in a boat and refastening it was called *suriage*. Replacing the bottom entirely he calls *sokoshinzo*.

Nasu said he only did suriage once or twice. He said only the lowest, central portion of the bottom was actually raised; the higher ends would have been less likely to have seen any damage from the boat hitting rocks or being dragged up on the shoreline. The nails at the chine had to be pulled out, the bottom lifted about .3 shaku, the side planking squeezed together and refastened. The excess material on the bottom of the dozuke was then cut off. He said he replaced bottoms

The two historic ubune in the collection of the Gifu City Museum of History, one built by Gosaku Ando in 1985.

several times, but confessed it was very difficult because the planking in the old days was thinner and more fragile. Typically a customer might come to him for a new bottom when the boat was about ten to fifteen years old, and with the repair the boat would last another ten years. At that point, he said, the boat would generally be replaced.

The most common repair was replacing individual bottom planks, which Nasu called *tsukesage*. Some customers with older boats would need two or three planks a year replaced to keep a boat going.

In recent times ubune last about ten to fifteen years and the bottom is definitely the weakest link in their construction. When we visited Seki to see the boats, most of which were less than a decade old, I noticed all the boats were showing signs of deterioration of the bottoms. Adachi, the eighteenth-generation usho from Seki who gave us our red pine, is notable for being the only usho active on the Nagaragawa whose boat does not have any fiberglass. His strict adherence to tradition may be the result of his status as having the longest lineage of any of the region's usho.

Today most usho eventually coat the entire bottom of their boats with a layer of FRP (fiberglass reinforced plastic). This provides both abrasion resistance, strength, and keeps the shiki watertight. The disadvantage of this material is that it adds weight. Much older ubune from Gifu can be found in Arashiyama, Kyoto, with the entire exterior, sides, and bottoms coated in fiberglass.

After seeing the boats in Seki I asked Nasu if he ever considered increasing the thickness of the bottom planking, but he repeated something he told us throughout the project: usho demand as light a boat as possible. In the old days the boats were even lighter, because fishermen traveled downriver and had to often haul them over shallows. Even today the usho have to pull their boats up onto beaches. Nasu also added, having all the hull planking the same thickness was more convenient for the boatbuilder, since he didn't have to worry about segregating material of more than one thickness for the sides and bottom. He also mentioned that all types of fishing boats got a little heavier as boats had to accommodate the loads produced by outboard motors.[27] In the end these construction details are a tradeoff between longevity and performance.

Looking inside one of the museum boats and the curious flared center plank of the shiki. This was not something Nasu did building his boats. Presumably his father didn't do this either.

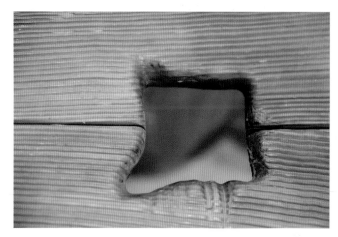

The ebisuana in one of the museum boats, worn into an interesting shape by the ropes run through it, reflecting a lifetime of hard use.

---

27   In Gifu the usho mount removable brackets and outboard motors on their boats. Prior to each evening's fishing, they motor upstream to wait for the sun to set and for the spectator boats to gather. They leave their motors on the beach and drift downstream fishing.

Although evoking ancient times, this scene of cormorant fishing in Ozu, Ehime Prefecture, is recent and the boats used are fiberglass. Photo courtesy of Kita Management.

Other Cormorant Fishing Boats

In 2018 I traveled throughout western Japan and China on a research trip funded by a grant from the Asian Cultural Council of New York City. This was a follow-up to my research in 2017 working with Seichii Nasu. My goal was to visit as many other of Japan's cormorant fishing sites as possible and survey the type of boats at each site. In most cases I was able to talk to fishermen and boatbuilders, and to measure the local boats. I then traveled to Guilin, China, and met the most famous cormorant fisherman there, who showed me his traditional raft and demonstrated how they are made. Below is a brief description and image of the variety of cormorant boats found in Japan and China.

Arashiyama, Kyoto

A handful of ubune work in this heavily touristed northwest corner of Kyoto on the banks of the Hozu River. My understanding is these are used boats purchased from Gifu, then entirely fiberglassed, inside and out. In the photograph, a single ubune is on the left and the remainder of the boats are spectator craft. The latter are very similar in design to the traditional work boats of the Hozu River.

Uji, Kyoto

Two ubune operate on the banks of the narrow Uji River, a short train ride south of Kyoto. In form the boats mimic the iconic design of the boat of Gifu, but sadly they are made of metal. Uji has become famous recently for having Japan's first female usho of the modern era, and a few years ago she took on another woman as an apprentice.

Miyoshi, Hiroshima

Miyoshi is a small city in the mountains in the center of Hiroshima Prefecture, located where the Gono, Basen, and Saijo Rivers merge. Cormorant fishing has a long history here, and like Gifu, the boats have remained unchanged for more than a century. Fishing for tourists has been done for the last one hundred years. The tourist boats are called *yuuransen,* which literally means "fun watching boat." They are 30.33 shaku long, built entirely of full-length Miyazaki cedar.[28] The ubune are striking: 30 shaku long and just over 3 shaku wide. In 2003 I met Mr. Kenji Mitsumori, the last professional builder of this region. His passing about ten

---

[28] Miyazaki Prefecture in Kyushu produces some of the finest cedar in Japan. It has historically been exported for boatbuilders as far away as Okinawa.

years ago precipitated a crisis, but Mr. Tenkyou Hirofumi, a house carpenter, came forward to study Mitsumori's boats and began to build them. He has to date built three ubune and three yuuransen. He now anticipates retiring soon and once again there is no successor. He mentioned he sometimes travels to Iwakuni to help another boatbuilder (see below).

Hirofuni and a city official took me to visit Mitsumori's widow, whom I had met in 2003. At the time of my earlier visit she was working alongside her husband. She said that her husband built boats for fifty years, and she estimated he built between 1,000 and 1,500 boats. When he was younger he could build a typical fishing boat in a week. His shop still had his tools and patterns. Tenkyou said he had modified Mitsumori's design slightly, thickening the bottom, which made the boat stiffer and easier to build, and letting the garboard planks run past the transom, which makes the boats track better.

Masuda, Shimane

Masuda is a small city at the mouth of the Takatsu River where it flows into the Sea of Japan. Cormorant fishing came very late here, spearheaded by one fisherman who was determined to establish the occupation. Eventually there were three fishermen using cormorants. Amazingly they did not leash their birds, but trained them to respond to voice commands. This is unknown elsewhere in Japan, although in China some fishermen use birds off leash. Also, these usho were not primarily fishing for tourists; they were commercial fishermen catching ayu with cormorants. The practice stopped about twenty years ago with the death of the senior usho. They did not use a specialized craft, just the local ayu boat used primarily for net fishing. These are of straightforward design, made of cedar, 21.4 shaku long. In Masuda they are called *hiratabune,* which means "flat-bottomed boat." The head of the local fisherman's union told me there are 800 registered fishermen on the Takatsu River, and while most fish with rods from the banks there are still a signifi-

cant number who net fish using wooden boats. However, the last boatbuilder is elderly and I could not meet him during my visit because he was ill. Traveling upriver I found a single fiberglass copy of these Masuda boats.

Iwakuni, Yamaguchi

Iwakuni is most famous for the Kintai Bridge, a stone and timber multi-arched span over the Nishiki River, near where it flows into the Inland Sea. Cormorant fishing takes place under the bridge seasonally for tourists. The boats are very reminiscent of the Gifu boats in shape, but they are shorter, made entirely of cedar, and have one less plank per side. After a typhoon destroyed several of the boats, the usho of Gifu donated some of their older boats to Iwakuni. The current boatbuilder, Toshio Hashimoto, is now 90. He thinks he is a third generation boatbuilder and his family always built sea boats exclusively. He and his father built 10 to 12 meter netfishing boats for the Inland Sea. Traditionally, building river boats and sea boats have been separate trades in Japan, but when the usho lost their boatbuilder about fifty years ago they approached Hashimoto to build ubune and viewing boats. Hashimoto is sometimes helped today by the boatbuilder in Miyoshi, Hiroshima (see above).

Ozu, Ehime

Ozu, a city on the Hiji River in Ehime Prefecture, has a long but interrupted history of cormorant fishing. A painting from the 1600s shows the local lord, or *daimyo,* watching cormorant fishing, but at some point the practice disappeared. Following the Word War II, in an effort to revive tourism, the city engaged with a Japanese who had lived in China before the war and watched cormorant fishing on the Yangtse. This man then purchased some birds in Kochi Prefecture, boats were built, and fishing commenced for tourists in 1957. Just as in Gifu, the height of this tourism in Ozu was in the 1970s with 28,000 tourists annually over a four-month season. At that time there were sixty tourist boats and three ubune. Photos show not a specialized watercraft but probably a version of the local small fishing boat. Today fifteen boats carry tourists who watch two usho in their boats, all of which are fiberglass. Ozu may be unique in having the tourist boats follow the fishermen over an extended stretch of the river (see pp. 17, 78).

Hita, Oita

The town of Hita lies along the Mikuma River in central Oita Prefecture. The town lost its professional boatbuilder about fifteen years ago, but a local truck driver named Matsumura took an interest in the boats and taught himself how to build them. They are a very straightforward design, 23 shaku long. One of the usho, Higashi Nishio, arrived and explained the reason for a small beam set in the bottom near the stern. He said when poling the boat one wedges a foot under the beam for stability.

Asakura, Fukuoka

Asakura is a small town twenty miles west of Hita on the Chikugo River. I had been told in Hita the former builder there also supplied the usho of Asakura, and indeed the boats were almost identical. The Asakura boats were slightly

longer at 23.6 shaku and the midship section slightly larger as well. The boats I saw were much older and therefore must have been built by Hita's former boatbuilder and not Matsumura. The other major difference was an arrangement built on top of the transom to attach an outboard motor.

Guilin, China

The Li River, flowing through the Pearl River Basin, is famous for the karst mountains with their sawtooth silhouettes. Cormorant fishing originated in China thousands of years ago and Guilin is considered the most famous site for this type of fishing. Today, however, cormorant fishermen merely pose for tourists. There is no fishing allowed on the Li at this time while efforts are underway to rehabilitate the water quality. The boats used by fishermen are bamboo rafts made by the fishermen themselves. Nearly identical rafts made of plastic pipe are common on the river. The fisherman I interviewed, who went by the name Blackbeard, showed me how the outer skin of the bamboo is stripped and then bent over an open fire. The most critical part of the process, he said, was oiling the finished raft and carefully allowing the bamboo to dry so it didn't split. I also met his brother-in-law, who demonstrated making a cormorant basket (as in Japan, the birds are taken to the boat in a special basket). Everyone acknowledged cormorant basketmaking was a much rarer craft than raft building.

Douglas Brooks teaching traditional methods of Japanese boatbuilding in 2022, above at the Japan House, University of Illinois Urbana-Champaign and opposite at the WoodenBoat School in Brooklin, Maine. Both classes built a 22-foot boat from the Shinano River in Niigata, Japan, used by farmers and fishermen for work, transportation, and dredging the extensive canals traversing the delta. Photos above by Fred Zwicky, University of Illinois; photos opposite by the WoodenBoat School.

About the Author

Douglas Brooks is a boatbuilder, writer, and researcher, specializing in the construction of traditional wooden boats for museums and private clients. In addition to building replicas of North American boats, Brooks has researched traditional Japanese boatbuilding since 1990, focusing on the design secrets and techniques of the craft. Brooks has apprenticed with nine boatbuilders throughout Japan since 1996. His boats have been exhibited at the Urayasu Folk History Museum, the Niigata Prefectural Museum of History, the Michinoku Traditional Wooden Boat Museum, Tokyo's Museum of Maritime Science, the Setouchi Festivale, and the Mizunoki Museum of Art. He has also built Japanese boats for the Peabody Essex Museum, Lowell's Boat Shop, Ritsurin Koen, Anderson Japanese Garden, and The Richard & Helen DeVos Japanese Garden. This is his fifth book on Japanese boatbuilding.

Brooks lives with his wife Catherine in Vergennes, Vermont. To see photographs of his work, please visit:
www.douglasbrooksboatbuilding.com

Praise for *Japanese Wooden Boatbuilding*

"A magnificent study. Its minuteness and clarity of detail will fascinate not only wooden boatbuilders, but anyone interested in the craftsmanship and culture of an earlier Japan."
— James F. English, Jr., Director Emeritus of Trinity College and Mystic Seaport Museum

"Reveals a world of sleek, practical forms perfected over centuries, of dedicated craftsmanship practiced today by just a few. Douglas Brooks's precious record of that world is dense with information, beautifully designed, and highly readable."
— Louise Cort, Curator, Smithsonian Institution

"A bible of the small wooden boat evolution."
— Dick Wagner, Founder, Center for Wooden Boats

"An insightful and enchanting window into the remarkable microcosm of Japan and its boatbuilding traditions."
— Tom Morse, *Ocean Navigator*

"An invaluable record of vanishing watercraft and the ways in which they are made, and a compelling record of a personal journey of discovery. Lavishly illustrated and lovingly wrought, it belongs on the shelf of anyone with more than a passing interest in the craft of wooden boat building. "
— John Summers, *WoodenBoat* magazine

Floating World Editions

Floating World Editions publishes books that contribute to a deeper understanding of Asian cultures. Editorial supervision: Ray Furse. Additional editorial support: Migiwa Imaishi. Book and cover design: Michelle Landry. Map by Benjamin Meader, Rhumb Line Maps. Printing and binding: Toppan-Leefung Printing Limited. The typefaces used are Garamond Premier Pro, Myriad Pro, The Sans, and Trajan Pro.